THE ETHNOBOTANY OF PRE-COLUMBIAN PERU

THE ETHNOBOTANY OF PRE-COLUMBIAN PERU

MARGARET A. TOWLE

AldineTransaction
A Division of Transaction Publishers
New Brunswick (U.S.A.) and London (U.K.)

First paperback printing 2007

This book is printed on acid-free paper that meets the American National Standard for Permanence of Paper for Printed Library Materials.

Library of Congress Catalog Number: 2006052289
ISBN: 978-0-202-30930-9
Printed in the United States of America

Library of Congress Cataloging-in-Publication Data

Towle, Margaret Ashley, 1903-
The ethnobotany of pre-Columbian Peru / Margaret A. Towle.
p. cm.
Reprint. Originally published: Chicago: Aldine, cl961, in series: Viking Fund publications in anthropology; no. 30.
Includes bibliographical references (p.).
ISBN 978-0-202-30930-9 (alk. paper)
1. incas-Ethnobotany. 2. Incas-Antiquities. 3. Ethnobotany-Peru. 4. Peru-Antiquities. I. Title.

F3429.3.E6T68 2007
985'.019—dc22 2006052289

ACKNOWLEDGMENTS

I WISH to express my sincere appreciation to Columbia University, the American Museum of Natural History, and the Peabody Museum of Archaeology and Ethnology of Harvard University, for making available the collections of archaeological materials upon which this study is largely based. Also, thanks are due to Mr. Louis M. Stumer of Lima, Peru, and to Dr. Alfred Kidder II of the University Museum, University of Pennsylvania, for additional plant specimens from Peruvian archaeological sites which served to increase the distribution of certain species.

I am indebted to Professor W. Duncan Strong of Columbia University, who first suggested a study of vegetal remains recovered from archaeological sites in Peru, and to Professor Paul C. Mangelsdorf, Director of the Botanical Museum of Harvard University, for their unfailing interest and encouragement. Professor Gordon R. Willey of Harvard University, Dr. Junius B. Bird of the American Museum of Natural History and Dr. Donald Collier of the Chicago Natural History Museum kindly read the manuscript and made helpful suggestions.

I also wish to express my appreciation for the generous advice given me by Dr. Albert F. Hill of the Botanical Museum of Harvard University on matters of botanical nomenclature as well as for his untiring editorial assistance.

It is a pleasure to express my gratitude to the other members of the Staff of the Botanical Museum, the Biological Laboratories, the Peabody Museum and the Harvard University Herbarium for their help and the many courtesies shown me.

Further, I am indebted to those who have assisted me in assembling the illustrative material: Mr. Elmer W. Smith who made the chart and the painstaking drawings of the archaeological plant remains; Dr. Samuel K. Lothrop and Mrs. Joy Mahler Lothrop of the Peabody Museum who made available specimens of Peruvian pottery for photographing; Dr. Alfred Kidder II and Mrs. Caroline Dosker of the University Museum, University of Pennsylvania, Dr. Junius B. Bird of the American Museum of Natural History, and the Director of the Chicago Natural History Museum, who supplied numerous photographs some of which are included herein.

To these and all others who have given so generously of their time and knowledge during the course of this study, I extend my sincere thanks.

TABLE OF CONTENTS

FOREWORD

A good many years ago, during my first visit to Peru as Duncan Strong's archaeological assistant, I was struck by the extraordinary state of preservation of peanut shells that came out of the middens at the famous site of Pachacamac. I think we commented at the time that the surface of the dig bore a strong resemblance to the main entry of Yankee Stadium after a Saturday game. In fact, as one accustomed to accepting crushed peanut husks between sidewalk cracks as part of the urban landscape, I did a "double-take" on this Pachacamac debris before being convinced that it was archaeological context and not something dropped casually by the workmen. But after a few weeks as an archaeologist on the Peruvian coast one becomes adjusted to the fact that "finds" from the dust-dry prehistoric sites of that rainless strip include much more than potsherds and objects of stone. Such traditional open-site "perishables" as cloth, wood, feathers, and plants simply did not perish—or at least perish completely or very easily—in that remarkable environment. For it is an environment that has preserved in millenia-deep refuse heaps a surprisingly full record of ancient life, and a portion of this record is that of the American Indian's development and utilization of cultivated plants—including the ubiquitous peanut.

The Ethnobotany of Pre-Columbian Peru, by Margaret A. Towle, is a work designed for both the archaeological and the ethnological specialist. The first will find in it much more than he is ordinarily conversant with in the way of botanical descriptions, identifications, and discussions; the second will see his data in proper cultural and chronological perspective. Third, the book is for any student of culture history or ethnobotany who is interested in the story of man and agriculture.

As we learn more about the rise and spread of native New World agriculture, it becomes evident that Peru was one of the hearths of its development. Plants were cultivated here at least 2,000 years before the beginning of the Christian era. Sedentary village life was intimately bound up with this cultivation, and later civilizations and empires rested upon it as a foundation. From Peru agriculture diffused to other parts of the Americas. We have yet a long way to go in tracing out the complex history of the interrelationships and exchanges of plant species between Peruvian highland and coast, between Peru and the American lowlands to the east, and between Peru and the other nuclear American center, Mesoamerica. But the present book is a good milestone on the road.

GORDON R. WILLEY
Harvard University

THE ETHNOBOTANY OF PRE-COLUMBIAN PERU

INTRODUCTION

One of the most interesting and in many instances rewarding aspects of archaeological research is the study and interpretation of the plant remains that have been recovered. The record of plants that have been used by man is of importance to both the botanist and the anthropologist. To the former it contributes information concerning the origin, history, distribution and uses of both wild and cultivated species. To the latter it adds knowledge of man's adjustment to his surroundings, and of the botanical aspects of his culture.

It is difficult to point to a single facet of man's life that has not in some way been associated with the plant world. Man's chief concern is in obtaining an adequate supply of his basic needs. These necessities consist of food, fuel, clothing, and shelter, all of which may be conditioned in varying degree by the climatic demands of his surroundings. Once these basic needs become satisfied, man, through accident, curiosity or ingenuity, will discover those plants that will furnish him with the tools and raw materials of industry. He will further isolate plants that will make life easier and in some instances satisfy his aesthetic needs. But his relationship to his botanical environment does not end here. It also finds expression in his art, religion, folklore, mythology and language. This all-pervading association has come to be known as ethnobotany, a term applied to the study of the relationship between man and the plant world, without limits to time or to the degree of his cultural development.

The data for formulating the ethnobotany of a culture group may be derived from several basic sources. Field studies on the utilization of plants, surveys of the botanical and ethnological literature of the area in question and the examination of plant remains from archaeological sites related to the group, are but three of the sources that may be drawn upon. The choice and extent of these will depend upon the scope and demands of the problem. The present study is an attempt to reconstruct within the limits of the available data the ethnobotany of the prehistoric cultures of the Central Andes. Since this subject is one concerned with a pre-Columbian culture area, the basic evidence must be derived primarily from archaeological sources. However, the resulting data will be further strengthened by historical, ethnological or botanical references whenever these are available. The ultimate value of ethnobotanical information lies in its relationship to other cultural aspects, as well as to the culture history, of the region as a whole. In the present study, sufficient data have been assembled to make possible such an interpretation. However, archaeological evidence imposes natural limitations upon the extent to which such relationships may be ascertained, and when the gaps cannot be filled from other sources, the conclusions must remain speculative.

Serious interest in Peruvian archaeology did not develop in the western hemisphere and abroad until the latter part of the last century, at which time European archaeologists and botanists took the more active part. However, within the last fifty years a number of important explorations have been conducted in the Central Andean region, some under the sponsorship of North American institutions and others under the able leadership of the Peruvians themselves. The number of reports on field work and the studies of various aspects of Peruvian archaeology increased in proportion to the amount of actual excavation carried on in this region. In the beginning, archaeological accounts were largely descriptive, with little emphasis on cultural relationships or continuity. In fact, the dating of pre-Columbian Andean cultures was largely a matter of guesswork. However, a change in the method of approach to Peruvian prehistory came in 1903, when the eminent archaeologist, Dr. Max Uhle, first called attention to stratigraphy in the burial area at Pachacamac, and in so doing, laid the foundation for stratigraphic research. The natural outgrowth of this was the establishment of a cultural chronology whereby archaeological materials could be described in terms of time as well as space. Although an absolute chronology for the Central Andean region is still wanting, archaeologists have gone far toward its formulation, and today they have at their disposal a fairly dependable relative chronology for the cultures represented.

In recent years, the explorations conducted in the Central Andean region have pointed more and more strongly to the latent importance of the area in American prehistory. It likewise became obvious that in order to obtain the best results from all these researches it would be necessary to coordinate the work of North and South American archaeologists. To meet this need, the Institute of Andean Research was founded in 1937 (Strong, 1943), and the explorations and research carried on under its sponsorship are largely responsible for the Central Andean region being one of the best known archaeological areas of the Americas today.

An important outgrowth of this coordinated approach has been the re-appraisal of various cultural aspects, as evidenced from archaeological materials, all of which contribute toward a clearer understanding of the cultures represented. In the light of this current trend, it has seemed advisable to make such a study of the plant remains recovered from archaeological sites and so to reconstruct the ethnobotany of these cultures, not only as to the utilization of both wild and cultivated plants, but in connection with the development of agriculture as well. From a survey of the possible sources for such a study, it was at once apparent that the collections made from sites stratigraphically excavated would furnish the core of the material, since it would be possible to place these plant remains within a chronological framework.

GEOGRAPHICAL SETTING

The great Andean chain that extends almost unbroken from the Caribbean to the Straits of Magellan is divided geographically into the Northern, Central and Southern Andes. Of these the Central Andes extend from northern Peru into Bolivia. This region not only forms a natural geographical entity, but it constitutes a cultural unit as well, characterized by a high type of development. Our knowledge of this cultural area is perhaps as great as of any in the New World. The Central Andes region is divided into three natural divisions: the *montaña* or forested areas, the highlands and the coastal desert strip along the Pacific. These divisions represent distinctive physiographic zones for both native and cultivated plants, and all have furnished habitable areas for man during his long sojourn in the Central Andes.

The Montaña. Although the Spanish word *montaña* ordinarily refers to mountains, in Peru it is used to designate the forested areas of the eastern slopes of the Andes and those that cover the highlands on the Ecuadorean border. The *montaña* of the eastern slopes consists of a tropical rain-forest which is a continuation of the jungle vegetation of the Amazon basin. This forest ascends the slopes to about 6,000 feet, where it gives way to a dense "mountain forest" of smaller trees. This transition zone, known as the *ceja de la montaña* or "eyebrows" of the *montaña,* extends to the tree-line at about 10,000–11,000 feet. However, in northern Peru this "mountain forest" continues across the Andes, which are lower in this area, and forms the forest covering of the highlands on the western slopes as well.

The Highlands. This division includes the main ranges of the Andes, the table lands between the deep canyons that cut the high-level surfaces, forming inter-Andean valleys, and the Western Andean slopes. The high-level table land is extensive and lies between 10,000 and 15,000 feet, the latter beyond the range of agriculture. Rising from these high-level plains are ranges of peaks often reaching a height of 18,000 to 20,000 feet. Cutting this highland surface are deep canyons through which flow the upper waters of tributaries of the Amazon. These rivers, as they reach lower altitudes, broaden into wide valleys, some of which, because of the more temperate climate, support extensive cultivated areas.

The Coast. The desert strip that parallels the coastal mountain range extends from Ecuador into Chile, its width depending upon the proximity of the rocky slopes to the sea. South of the Piura River in northern Peru, there is a wide area of lowlands, the greater part of which is moving sand dunes. Slightly farther south, rocky slopes rise directly from the ocean, with a few small, protected, rock-bound harbors. From the Pativilca to the Pisco rivers, the alluvial soil brought down to the coast by the many streams forms large fans that appear to join to form a narrow lowland. This area is occasionally traversed by rocky spurs from the Andean range. Between the coastal range and the base of the Andes is a bleak, rocky terrain that begins to rise a little below the top of the coastal range and gradually increases in altitude to 5,000 feet at the lower Andean slopes

The desert strip is crossed by fifty-two rivers of varying size (fig. 1), only ten of which have sufficient water during the entire year to continue across the desert and empty into the ocean. However, the marginal lands of these streams support cultivated fields with the aid of irrigation as they did in prehistoric times. Because of the aridity and ruggedness of the coastal terrain, there are few locations, with the exception of these river valleys, that lend themselves to agricultural activities.

The climate of the Peruvian coast is both cool and dry. It is controlled by a combination of steep mountains, the Humboldt Current and the prevailing south and southwest winds. The climate of the greater part of South America is dependent upon the eastern trade winds that blow over the Amazon Basin where they become saturated with moisture. This is deposited in the form of mist and rain on the eastern slopes of the Andes, thus leaving the western Andean slopes entirely dependent upon the climatic conditions originating in the Pacific. Here the cold Humboldt Current, flowing northward from southern Chile and parallel to the Peruvian Coast, causes the humid south and southwest winds to lose much of their moisture at sea. This mass of cooled air gives rise to fogs which are particularly heavy during the winter months. Upon reaching the coast, the upward currents are not sufficiently strong to carry this cool moist air to an altitude which would lower the temperature to that of condensation. As a result coastal Peru receives but little rain. In fact, some sections receive rainfall only once in several years. This lack of precipitation on the coast increases the necessity for irrigation in the narrow, fertile river valleys. At certain times of the year, however, the heavy clouds and fog resting on the low foothills and slopes from 1,000 to 5,000 feet soak the ground sufficiently to cause a dense seasonal vegetation known as the *loma* to flourish. This supplies pasturage from June to October.

CULTURAL SETTING

The central section of the Andes possessed not only a geographical unity but a cultural one as well. Many of the coastal river valleys, the larger highland basins and the plateau south of Lake Titicaca, supported populations of varying size and complexity during the long period of prehistoric cultural growth. Although these separate areas may be described as individual entities, archaeologists usually group them into six geographical subdivisions, three of which are ascribed to the coast and three to the highlands. These subdivisions are not arbitrary but present well-defined cultural characteristics that set them apart and thus give a basis for comparisons on a large scale (Bennett, 1948).

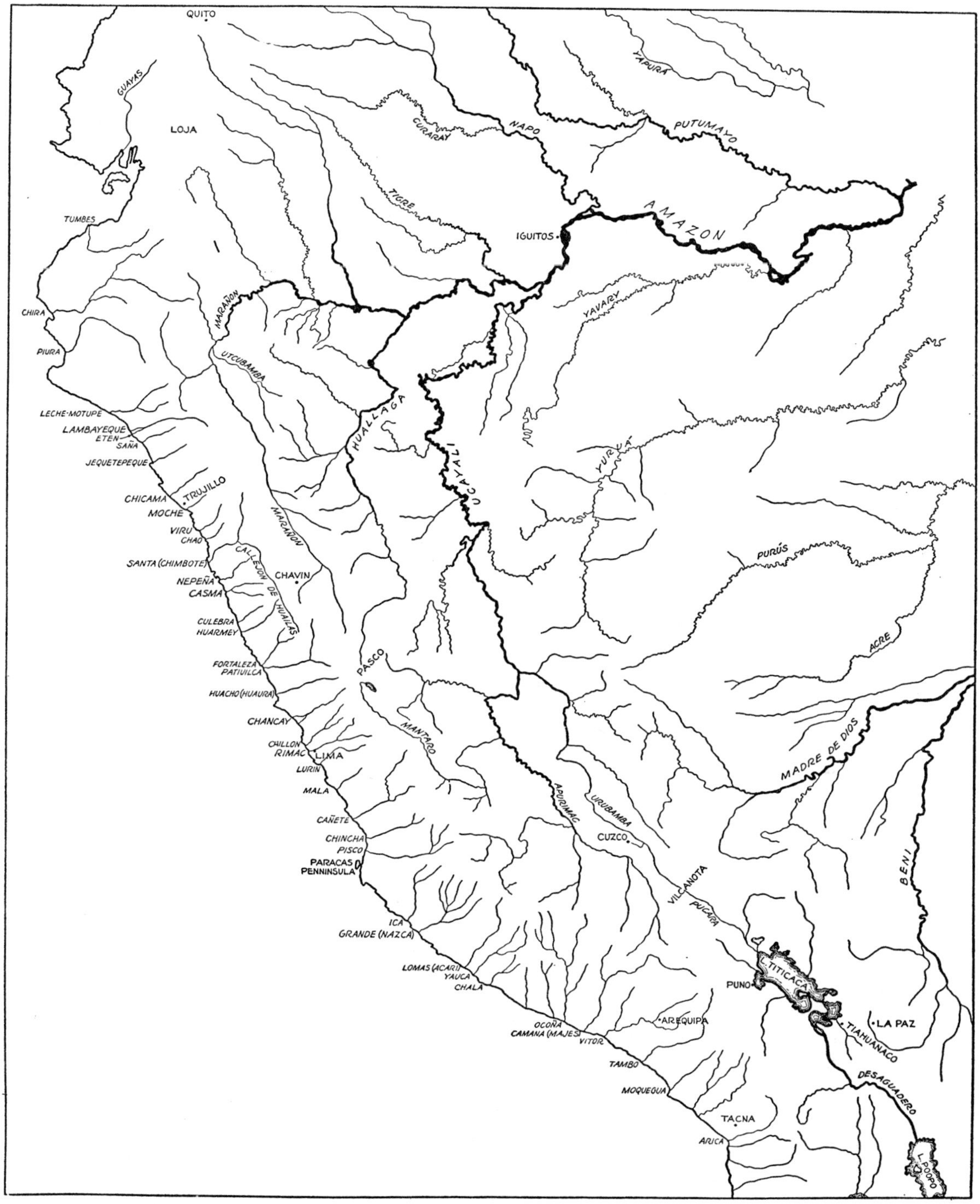

FIGURE 1
River Valleys of the Central Andes

These basic cultural characteristics, or traditions, covering a long period of time and space, form the structure for the unity of Central Andean culture.

The three coastal subdivisions are designated as the North, Central and South Coasts. Each division consists of a number of river valleys, which may be grouped together on the basis of similarities of archaeological materials that have been recovered (see also Bennett and Bird, 1949).

The North Coast. This division includes all the river valleys from Piura to Casma. Although archaeological work has been conducted in many of these, the cultural history of the Chicama, Moche and Virú is best known. Here the numerous sites have been thoroughly and systematically studied, the results possibly constituting the most complete developmental sequence of any of the Peruvian coastal areas.

The Central Coast. The region from the Huarmey to the Lurín Valley is described as the Central Coast. Here the most dependable information has been collected from the valleys of Chancay, Supe, Ancón, Rimac and Lurín. The Ancón and Lurín contain the well-known sites of Ancón and Pachacamac.

The South Coast. This division comprises the valleys from the Rio Mala to the Rio Lomas, a region which includes the archaeologically important Ica and Rio Grande de Nazca Valleys and the peninsula of Paracas. It is from sites located here that the famous South Coast textiles and pottery have largely been obtained. Recently Strong (1954; 1957) conducted extensive research in the area to collect data for the establishment of the cultural sequence represented.

The highlands of the Central Andes are likewise divided into North, Central and South subdivisions.

The North Highlands. This region begins at the Ecuadorian border and extends to the Department of Huanuco. The archaeological data for this area are limited principally to the important sites of Callejón de Huaylas, Huamachuco and Chavín de Huántar.

The Central Highlands. This designation includes the area from Huanuco to Cuzco. Our information is restricted mainly to the Inca capital of Cuzco and to certain early sites east of Lima, in the Rio Mantaro Valley.

The South Highlands. This extensive area comprises the plateau of southern Peru and the altiplano of Bolivia. Archaeological research has been mostly confined to sites in the region about Lake Titicaca.

These six cultural-geographical subdivisions supply the dimensions of space to Central Andean culture history.

CHRONOLOGY

A second dimension necessary to historical-cultural reconstruction, that of time-depth, is furnished by a relative cultural chronology established for the region as a whole (fig. 2). The source of the essential data upon which this chronology has been built has been derived from numerous stratigraphically excavated sites at which, in the majority of cases, local cultural sequences have been determined. These have then been compared with still other sequences established for sites in the immediate area in an attempt to trace the extent of the distinguishing characteristics insofar as the existing evidence will allow. The relationships so determined thus serve to connect the purely local sites with the region as a whole.

From comparative studies, it has become apparent that the cultural sequences of each geographical subdivision are marked by well-defined developmental *periods,* which act as time markers, each of which is characterized by distinctive manifestations. These rather numerous time markers have been variously designated by names derived from the type site at which the representative culture was first observed, by local geographical names or even by terms descriptive of the dominant ceramic style. The periods in turn have been further grouped into larger units in the time scale, designated as *epochs* or *stages,* which are based on the developmental processes demonstrated in the component periods; as with the periods, the lines of separation between them must remain, at least for the present, arbitrary. There is no standard terminology for these cultural epochs. In the present work, they are designated as *Pre-Agriculture, Incipient Agriculture, Formative, Classic, Fusion, Kingdoms and Confederacies, Imperial* and *Colonial.*

Since no calendar system or written records exist for the cultures that flourished in the Central Andes, the chronology that has been formulated is relative and not absolute. In order to give some approximation of the extent of the periods and the stages of this developmental sequence, the use of "guess dates" has been resorted to. This has been the accepted method in the past of treating the element of time, and so long as it is remembered that it is relative and not exact, it is useful. With the recent discovery of carbon 14 as a method for determining the age of materials, the possibilities for more accurate, if not exact, dating of specimens and the culture that they represent has been greatly enhanced. Many of the archaeological specimens that have been tested have been found consistent with the ages obtained earlier by the more laborious comparative methods of archaeological technique (Bird, 1951). As the carbon 14 method becomes further perfected, such approximate dates may be revised and a more dependable framework of time depth will result.

The tracing of cultural characteristics in the Central Andes has been greatly facilitated by the existence of *horizons,* each of which is defined on the basis of one or more art styles or techniques, or in some cases by a complex of cultural elements. These horizons vary in their age as well as in their extent. The six most important and widespread, presented in their chronological sequence, are *Chavín, White-on-Red, Negative Painting, Tiahuanaco, Black-White-Red* and *Inca.* Of these, Chavín, Tiahuanaco and Inca are of highland origin.

DATE	EPOCH	COAST: NORTH	COAST: CENTRAL	COAST: SOUTH	HIGHLANDS: NORTH	HIGHLANDS: CENTRAL	HIGHLANDS: SOUTH	DATE
	COLONIAL							
1532 1000	IMPERIAL KINGDOMS & CONFEDERACIES FUSION	Inca Chimu & La Plata Black-White-Red Coast Tiahuanaco	Inca Chancay & Black-on-White Coast Tiahuanaco	Inca Late Ica Middle Ica Early Ica Coast Tiahuanaco	Inca Local Tiahuanaco	Inca Early Inca Huari	Inca Local cultures Decadent Tiahuanaco	1532 1000
500 ↑ A.D.	CLASSIC	Huancaco Mochica Gallinazo III Gallinazo II	Nievería Maranga	Nazca	Recuay	Waru Derived Chanapata	Classic Tiahuanaco Pucara Early Tiahuanaco	500 ↑ A.D.
B.C. ↓ 500 750	FORMATIVE	Gallinazo I Salinar & Puerto Moorin Cupisnique	Playa Grande Baños de Boza Early Ancón	Paracas Cerillos	Huaraz W-on-R Chavín de Huantar	Chanapata	Chiripa	B.C. ↓ 500 750
1200	INCIPIENT AGRICULTURE — INITIAL CERAMIC	Middle Guañape Early Guañape	↑ Aspero ?					1200
1500 2000 2500	INCIPIENT AGRICULTURE — PRECERAMIC	↑ Huaca Prieta & Cerro Prieto		Lomas				1500 2000 2500
↓ 8000	PRE-AGRICULTURE			San Nicolas				↓ 8000

FIGURE 2
Chronological Chart for the Central Andes

The Chavín horizon has been described on the basis of a stylized feline motif, which was used on various media. Although evidence of this horizon has been recovered from the North Highlands and the North, Central and South Coasts, the exact center of dispersal is obscure. The White-on-Red Horizon, which is found on the Central and North Coasts and in the North Highlands, is based mainly upon a technique of ceramic painting in which designs in white are applied by brush on the natural red clay of the vessel itself. Likewise, the Negative Painting Horizon is determined by a ceramic technique in which the pattern is applied with some removable wax-like resistant or other substance such as clay. The whole vessel is then treated, after which the substance is removed, leaving the motif in the original color of the pottery. This horizon has been found in the North Highlands and on the North, Central and South Coasts.

In contrast to the above, the Tiahuanaco horizon consists of a cultural complex including an art style, distinctive ceramic painting and pottery shapes. This horizon is pan-Andean in distribution, being found throughout the highland and coastal regions. However, the exact center of its dispersal is unknown. The Black-White-Red is dominated by three-color ceramic painting and the use of a distinctive flask shape as well. Sites in the North Highlands and on the North, Central and South Coasts contain representative specimens from this horizon. The most recent of these widespread cultural influences is the Inca horizon, which like Tiahuanaco is pan-Andean in its distribution. However, this horizon is known to have spread from the area about Cuzco, the ancient Incan capital. It is represented by a complex of cultural elements including an art style and characteristic pottery shapes.

On the chart (fig. 2), the Central Andean area has been divided into the six cultural-geographical divisions of the region and the cultural sequence for each of these has been indicated.

MATERIALS AND METHODS

Through the kindness of Dr. William Duncan Strong of Columbia University, the writer was given the opportunity of examining plant specimens from various sites on the North, Central and South Coasts of Peru. These collections comprise approximately 2200 specimens, representing both wild and cultivated species; some of the sites from which they were recovered represent the earliest pre-agricultural epoch, while the upper levels of still others date from the time of the Conquistadores. The individual chronologies of these sites with their plant distributions furnish comparisons in time as well as space. Representing the North Coast of Peru we have a collection of some two hundred plant specimens from sites located in the Virú Valley, excavated in 1946 by the Columbia University Unit of the extensive Virú Valley Project. A monograph describing the plant material appears as Appendix 2 of *Cultural Stratigraphy in the Virú Valley* (Strong and Evans, 1952).

The Central Coast collection consists of two groups of material: one from the mouth of the Supe Valley, the other from the extensive site of Pachacamac, located 30 kilometers south of Lima. The specimens from the Supe Valley were obtained at the Lighthouse and Aspero sites during the excavations of Project 3 of the Institute of Andean Research during 1941-1942. The collection is composed of approximately ninety specimens, the majority of which were published in *Early Ancón and Early Supe Cultures* (Willey and Corbett, 1954). The collection from Pachacamac was also made in 1941–1942 under the sponsorship of the Institute. For the purpose of establishing a ceramic sequence based upon stratigraphy, two cuts were made in the large midden below and to the south of the main terrace of the Temple of the Sun. A report (Towle, 1948) on the plant remains collected, which comprise about seven hundred specimens, although completed, has not yet been published. Additional material, comprising 125 specimens from recently excavated sites in the Rimac Valley, made available by Louis M. Stumer of Lima, has also been studied (Towle, 1958).

The material from the South Coast was obtained during the Columbia University Expedition of 1952–1953 and represents nine sites in the Ica and Nazca Valleys. This is the largest collection, comprising nearly a thousand specimens. This study (Towle, 1956) has been completed and will appear as part of the report of the expedition. In addition, through the courtesy of Dr. J. O. Brew, Director of the Peabody Museum, Harvard University, the writer was given the opportunity of examining the plant remains in a mummy bundle found at Paracas Necropolis. These specimens are described in *Plant Remains from a Peruvian Mummy Bundle* (Towle, 1952b). Still other material from the South Coast was made available for comparison by both the Peabody Museum and the American Museum of Natural History in New York. Through the courtesy of Dr. Alfred Kidder II, of the University Museum, University of Pennsylvania, the writer has had the opportunity of examining the small but unique collection of plant remains obtained at the site of Chiripa in the south highlands.

With few exceptions, the collections examined consisted of unworked plant materials. Specimens of manufactured articles of plant origin such as textiles, basketry, matting, gourd vessels, wooden objects, etc., have usually been the subject of separate studies. However, a few articles showing workmanship were included. Among these were beads made of seeds, some gourd containers, matting and rope, and certain manufactured articles that were included in the mummy bundle from Paracas Necropolis.

Recently it was the author's good fortune to be given by Dr. Junius B. Bird the opportunity of studying the extensive collection of vegetal remains from Huaca Prieta de Chicama. It has been possible to include the results of the study of several species in the present work. However, a comprehensive report on the plant remains from this important site is being prepared in collaboration with Dr. Bird.

The earlier archaeological literature contains many references to vegetal material from sites that were excavated without the benefit of stratigraphy. Consequently, in most cases, these specimens cannot be definitely assigned to their proper cultural provenience. Nevertheless, since much of this material has been identified, our knowledge of the number of species used by the pre-Columbian Peruvians has been increased. To this group of literary references belong the reports written on excavations made in Peru in the latter part of the nineteenth century. During this period of awakened interest in Peruvian prehistory, not only archaeologists but botanists as well were aware of the importance of the discoveries being made. This latter group of scientists, stimulated by the work of DeCandolle and others on the origin of cultivated plants, saw in the plant remains from Peru a source of possible evidence for agreement with or refutation of the current theories. This interest resulted in the description of plant material found in archaeological sites and in papers devoted to such material and its interpretation *per se*.

With the exception of a few short and scattered references, the first report describing Peruvian archaeological plant remains in any detail appeared in 1876. Dr. Saffray, in his *Les antiquités péruviennes à l'exposition de Philadelphie* (1876), carefully describes a mummy bundle included in that exhibit, although the site from which it came is not mentioned. He attempts to identify the plant materials that were represented, both as worked and unworked remains, using for reference, in some instances, certain ethnological accounts of South American tribes that were then available. In 1879, the botanist, Alphonse Tremeau de Rochebrune, published an article entitled *Recherches d'éthnographie botanique sur la flore des sepultures péruviennes d'Ancón.* This paper was a report on the author's examination of plant specimens from Ancón included in the collections made by M. de Cessac and Dr. Ludovic Savatier. Rochebrune describes these remains under two categories: *Food and Medicinal Plants* and *Industrial Plants and Dyes*. It is unfortunate that the author in many instances does not mention what part of the plant he analyzed, or in some cases, whether he based his study upon the finished article or on a fragment of the raw material of which it is made. Such information is of great importance in determining the reliability of any identification.

During the years 1880–1887, W. Reiss and A. Stübel published the report of their excavations at Ancón, *The Necropolis of Ancón in Peru,* one section of which ("Plants and Fruits," vol. 3, chap. 13) is devoted to descriptions with identifications of the plant materials that were recovered. Accompanying this chapter are three excellent plates that clearly illustrate the text. This study was prepared by the German botanist, L. von Wittmack. Later, Wittmack published *Die Nutzpflanzen der alten Peruaner* (1888), in which he discusses the plant material recovered from archaeological sites in Peru more fully than in his previous work. He draws not only upon the Ancón material found by Reiss and Stübel, but also upon the collections of Commander Acland and Lieutenant

Holland, then housed in the Museum at Oxford. The "Moseley Collection," consisting of four mummies, was obtained by Commander Acland from Ancón, while the place of origin of the Holland specimens is unknown. Like Rochebrune, whose work on the Peruvian plant remains is referred to by Wittmack, the latter also treats the subject on the basis of economic use, describing the species under eight headings: *Grains and Breadfood, Legumes, Tubers, Fruits, Vegetables, Narcotic Plants, Spices and Medicinal Plants,* and *Technical Plants.*

Additional plant material discovered by Captain Berthon at Pachacamac, Chorrillos, Ancón and La Rinconada, a site near Lima, were identified by J. Costantin and D. Bois, the results appearing in their paper, *Sur les graines et tubercules des tombeaux péruviens de la Periode Incasique* (1910). This report contains careful descriptions of the specimens, several of the species identified not having previously been reported from Peruvian sites. In 1917 William E. Safford published *Food-Plants and Textiles of Ancient America,* a study based upon collections made by the author in Chile, Peru, Bolivia, Argentina and various parts of the United States. Included are a number of references to species used by the people of prehistoric Peru which were identified from actual remains or from depictions of plants in art.

A re-appraisal of these earlier ethnobotanical works was given by Hermann von Harms in his comprehensive study, *Ubersicht der bisher in altperuanischen Gräbern gefundenen Pflanzenreste* (1922). In addition to the already published reports on the identification of Peruvian plant remains, this well-known German botanist included the results of his own studies of specimens recovered from Chuquitanta, Pachacamac, Ancón and the Department of Ica. This work constitutes a valuable contribution to the field of prehistoric Peruvian ethnobotany, for not only does Harms present his new material but he also carefully checked the identifications of many of the specimens deposited in the museums of Berlin which had been described by Rochebrune and Wittmack. This scholarly paper has been of particular value in bringing up to date many of the botanical names and in helping to clarify some of the inconsistencies found in the earlier studies.

The most recent work devoted to the plants used in pre-Columbian Peru is that of Eugenio Yacovleff and Fortunato L. Herrera, *El mundo vegetal de los antiguos peruanos* (1934–1935). While the earlier sources are concerned only with archaeological material, these authors based their study upon the identification of the plants described by the chroniclers of the Conquest and Colonial periods. Many of the species discussed, however, have not as yet been recovered from archaeological sites. Their descriptions of the various plants are often substantially enriched by added information from ethnological and archaeological sources. The latter take the form of references to depictions in art, to actual plant remains now in the museums of Peru, and in some cases to the earlier writings on prehistoric vegetal specimens mentioned above. However, only those species with archaeological associations have been utilized in the present study.

All of the references either to actual archaeological plant remains or to plants depicted in art contribute to the reconstruction of the ethnobotany of the pre-Columbian cultures of the Central Andes, which has been found to reflect in varying degrees the general cultural history of the region.

The plant remains which were examined by the writer were, as a whole, fairly well preserved, due chiefly to the arid conditions of the desert coastal strip in which they were found. Many of the specimens, such as maize and gourds, retained their identity, and usually the genus and often the species could be determined. Other specimens, although well preserved, were too small or too lacking in the necessary details to make identification possible. Still other materials were of finer texture, the more common examples of which were leaves and stems. Such remains were usually fragmentary, a condition that often made identification difficult if not impossible. Regardless of the condition of the specimens when recovered, the methods used in identification relied primarily upon a study of the gross morphology of the specimen and microscopic examination, as well as upon comparison with both herbarium material and archaeological specimens which had previously been identified. Floras and botanical monographs and historical and ethnological accounts of the Central Andean region were freely drawn upon.

The species represented by the actual archaeological materials examined by the writer and those referred to in the other source material are described in Part I. These data are further developed in Part II for each of the six geographical-cultural subdivisions of the Central Andes in terms of their respective cultural chronologies. Emphasis has been placed upon the cultural associations of the remains; the culture fabric of which they were a part has been briefly outlined. Particular attention has been given to the trends in the use of wild and domesticated species and the development of agricultural practices. Wherever possible the use of plants and the development of agriculture in the different areas have been compared. Finally, the ethnobotanical data and the related agricultural practices are summarized for the central Andean region of Peru as a whole in order to emphasize the evolutionary trends that existed.

PART I

THE ETHNOBOTANY OF PRE-COLUMBIAN PERU

In our discussion of the ethnobotany of pre-Columbian Peru, the various species are considered by families following the Engler and Prantl system of plant classification, with modifications; in each family the genera and species are arranged in alphabetical order. The species name is followed by full citation of authorities and place of publication. Only the synonyms essential to the present work have been included, however. For the purpose of minimizing any confusion that might arise, there are also listed those incorrect names that have appeared in the earlier literature.

Following the synonyms, the more commonly occurring vernacular names are given. These do not profess to be complete, nor do they include any minor variations in spelling. A plant may have several common names, each derived from a different locality, but often it is not possible to tell to which location the name belongs. In the text the names common to the Peruvian coastal region, whenever so designated, have been used, whereas in those instances where the locality is not specified, the terminology most used in the literature has been selected. It should be noted here that a name may refer to plants of more than one species of a genus or even to species of different genera when the plants have a similar appearance. The common name *huarango* will illustrate the former; it is used to designate the two leguminous species, *Acacia macrocantha* and *Prosopis chilensis.*

A short description of each species is furnished to give readers not familiar with the plant some general idea of its appearance and its more marked characteristics. In the case of the less well known species, parts of the plant which might be found among archaeological material are described in greater detail. Such information, it is hoped, will aid those attempting to identify similar vegetal remains.

Following the botanical descriptions, mention is made of whatever archaeological remains have been recorded. An enumeration of various specimens that have been found is given, together with the name of the site and the bibliographical reference. Part I does not purport to be a catalogue of all the plant material that has been found archaeologically, nor of all the representations in prehistoric art. As stated above, it is primarily a study based only upon the collections examined by the writer and upon the archaeological data in the papers listed in the bibliography. The specimen itself, the site where it was procured and whenever possible, its cultural provenience, constitute the basic data with which we are concerned.

CRYPTOGAMAE

Pieces of the small stalks of a sea alga were found in association with cotton in a Paracas mummy bundle (Yacovleff and Muelle, 1934, p. 134).

LAMINARIACEAE. Laminaria Family.

MACROCYSTIS *Ag.*

Macrocystis Humboldtii *Ag.* Kunth Syn. I (1822) 6
Aracanto (Yacovleff and Herrera, 1935, p. 74).—*Sargoso* (*ibid.*).

This giant species of the brown algae is found in the Pacific waters of the Peruvian coast. The plant grows from a ramifying basal system, the long branching stalks attaining great length. These branches bear leaflike structures, *laminae*, each of which has a small, gas-filled bladder at its base. We have no information regarding the uses of marine plants in pre-Columbian times. However, a Nazca vase recovered from a burial has a painted design composed of fishes and the long trailing stem of a seaweed. This latter has been identified as that of *Macrocystis Humboldtii* (*ibid.*, fig. 61).

POLYPODIACEAE. Fern Family.

The *Polypodiaceae*, consisting of about 170 genera, constitutes the largest family of ferns, and one which is widely distributed, particularly in humid or moist regions.

POLYPODIUM *L.*

This is a large genus of about 200 species growing in various habitats in the temperate and tropical parts of both hemispheres. The leaves are either simple or compound with large or small, naked sori located on the back of the veins or on the margins of the leaf. A number of species of *Polypodium* are to be found in the area of Cuzco (Herrera, 1941b).

The large, handled jars or *aryballos*, one of the most striking pottery types made by the Incas, are often decorated with stylized fronds of a fern placed vertically around the bowl of the vessel (pl. VII, A). These leaves, with clearly delineated veins and sori, resemble those of species of *Polypodium* (Tryon, 1959).

Among the Quechua, the word *r'aqui*, used either singly or as *r'aqui r'aqui* to designate various ferns, is also applied to this type of vessel. The presence of the ferns on the *aryballos* in all probability suggested the word as an appropriate one for the jar itself.

PHANEROGAMAE

TYPHACEAE. Cat-tail Family.

TYPHA (*Tourn.*) L.

Typha is a genus of about ten species which grow in swampy areas of the temperate and tropical regions of the world.

Typha angustifolia *L.* Sp. Pl. (1753) 971.
Typha domingensis Kunth Enum. Pl. 3 (1841) 92.
Cat-tail.—*Enea* (Herrera, 1939, p. 178).—*Totora,* in the Argentine (Parodi, 1932, p. 3).

The cat-tail is found in moist locations of the coastal valleys. Its strong stems, which are sheathed in long, sword-shaped leaves, rise from stout, scaly-leaved rhizomes. Both the fibrous stems and leaves lend themselves readily to the construction of matting, baskets and ropes,[1] while the rhizomes have been reported as a source of food.

Specimens of *Typha* have been recorded from archaeological sites in the northern, central and southern coastal areas. Bird (Bennett and Bird, 1949, p. 120) reports finding the rhizomes, which had presumably been used for food, in the pre-ceramic refuse at Huaca Prieta. At Huaca de la Cruz in the Virú Valley, two ropes have been found, one of two-ply, the other of four, made of the twisted leaves of this plant (Towle, 1952c, p. 352). Both specimens came from Mochica levels. Farther south on the central coast, at the Aspero site at Supé, gourd bottles still containing plugs of folded cat-tail leaves were found in graves (Towle, 1954, p. 130). Also from the same site, and likewise from a grave, a large mat was recovered whose perpendicular strands have been identified as the leaves of *Typha* (*ibid.*, p. 130). At Ancón, a mat used to cover a mummified dog was constructed of these leaves (Wittmack, 1888, p. 347). The mid-section of three-piece or "tripart" spindles found at sites on the Central Coast also were made from sections of the flowering stalks of the cat-tail, as were the heddle pattern rods of some looms recovered there (Bird correspondence, 1957). Stems of the cat-tail were used in the construction of many of the large baskets which contained the heavily swathed mummy bundles found at Paracas Necropolis (Yacovleff and Herrera, 1934, p. 294). In addition, this plant is said to have contributed material used in the embalming of the mummies, but neither the part of the plant nor the exact purpose for which it was used is specified (*ibid.*).

[1] Much confusion exists in both the archaeological and ethnological literature as to the identity of the materials used on the Peruvian coast in the manufacture of matting, basketry, cordage, "reed" rafts and small boats. *Typha* and also certain species of *Cyperus* and *Scirpus* were employed in their construction, but unfortunately the common names applied to them are used interchangeably. The difficulty encountered in the literature appears to arise when the common names are the only basis for identification.

GRAMINEAE. Grass Family.

CALAMAGROSTIS *Adans.*

Calamagrostis, a large, widely distributed genus, consists of perennial grasses with unusually long, narrow blades and strong, thin culms, which vary in height from twenty to 100 cm. Macbride (1936) reports a wide distribution for the genus in Peru and lists twenty-five species.

Calamagrostis spp.

Specimens attributable to this genus were found among the plant remains from Pachacamac (Towle, 1948). Because of the poor condition of the inflorescence, it was not possible to identify the species. Another specimen referable to *Calamagrostis* was collected by Bird at Huaca Prieta where it is said to be used by the natives for making brooms (Bird correspondence, 1948).

CHUSQUEA *Kunth*

Chusquea, a genus of bamboo, occurs in Mexico, the West Indies and South America, where it is especially abundant in the Andes.

Chusquea scandens *Kunth* Pl. Aequin. 1 (1822) 254.
Kurcur, Cuzco (Herrera, 1939, p. 216).

This plant, with solid, pliable stems, grows erect to a height of five meters or trails over trees and shrubs. The stems are used locally for making baskets. The plant is characteristic of the mountain forest regions from Bolivia to Colombia, and grows at altitudes from 600 to 3,000 meters.

Yacovleff and Herrera (1934, p. 264) state that this species may be the one referred to by the early chroniclers as the source of the pliable stems used by the prehistoric inhabitants of Peru for making jackets, helmets and shields. They also suggest the possibility that the plant is the one represented in the pottery of Moche. However, there are other large monocotyledonous plants which are indigenous to the coastal area and these might better have been the inspiration for the clay reproductions. Rowe (1946, p. 243) states that there is much basketry, cordage and matting in the tombs near Cuzco. Few of these have been studied, but it would appear that the Inca made technically excellent examples of this type of material. It is possible that the genus *Chusquea* is represented among these specimens.

GUADUA *Kunth*

Guadua, a genus of twenty species, occurs in the tropical and subtropical regions of Mexico and South America.

Guadua angustifolia *Kunth* Syn. Pl. 1 (1822) 253.
Caña de Guayaquil (Yacovleff and Herrera, 1935, p. 34).—*Ipa* (Macbride, 1936, pt. 1, p. 109).

This species is a stout bamboo, often attaining a height of ten meters or more, with stems fifteen cm. in diameter and relatively narrow leaf-blades, 1.2 to 1.8 cm. wide and about fifteen cm. in length. Like other bamboos, it is used by the natives for construction purposes. Drinkable water may be obtained from the nodes of the stems. Although *caña de Guayaquil* has been reported from Lima (Macbride, *ibid.*), its distributional range is in the tropical areas of the Amazon basin and from Colombia to the Guianas. Yacovleff and Herrera (1935, p. 34) identify the plant mentioned in early accounts of the ritual employed in certain burial customs of Colombia as *Guadua angustifolia.* The ritual consisted of placing one or more canes in the grave in a vertical position at the sides and the head of the body. These authors note the similarity between this ritual and the remains of what appears to have been a similar ceremony in the graves at Nazca. A further reference from the early accounts is to the use of the stems of a plant, again identified as *caña de Guayaquil,* for rowing rafts, a use which has been confirmed by representations on pre-Colombian pottery (Yacovleff and Herrera, 1934, fig. 21).

GYNERIUM *H. and B.*

Gynerium, a genus of three species, is native to tropical and subtropical America.

Gynerium sagittatum *(Aubl.) Beauv.* Ess. Agrost. (1812) 138.
Caña brava (Herrera, 1939, p. 133).—*Pintoc,* Cuzco *(ibid.,* p. 279).

This large grass is characteristic of swampy areas, where it often forms dense stands in the lowlands as much as ten meters in height. The sterile culms are crowned by a large, fan-shaped leaf-cluster, while others support the long, plume-like panicles. The edges of the leaves are slightly serrated, one of the characteristics used in their identification.

The tough, durable culms of *caña brava* are used extensively today for both construction and manufacturing purposes, and it is therefore not surprising that the archaeological evidence shows that it was similarly utilized in prehistoric times. Worked as well as unworked specimens have been recovered from various archaeological sites, and the frequently-encountered manufactured artifacts include baskets, containers, arrows, etc. (Yacovleff and Herrera, 1934, p. 262). Large fragments of a section of a culm of this grass came from the Gallinazo site of Castillo de Tomaval in the Virú Valley (Towle, 1952c, p. 352). Again, at another site (Huaca de la Cruz) in this valley, a disk was found, presumably a cover, made of the folded leaf or leaves of this species. This specimen was in association with Mochica elements *(ibid.).* Yacovleff and Herrera (1934, p. 263) mention the finding of mummy baskets at Paracas Necropolis made of the fibrous leaves of *caña brava.* In addition, there were specimens of armor, arrows, combs and other small objects, constructed from the tough stems of the plant. These authors further mention finding at sites in the Chillón valley stems of *Gynerium* two to four meters high with an attachment at the end which suggested the plume-like inflorescence. The purpose of these is not known, but the authors think that there is a possibility of their having been used as "banners" or standards.

PHRAGMITES *Adans.*

The genus *Phragmites* consists of three species, one from Asia, one from Argentina and the "marsh" or "reed grass" which occurs in all parts of the world.

Phragmites communis *Trin.* Fund. Agrost. (1820) 134.
Common marsh grass, reed.—*Carrizo,* Lima (Herrera, 1939, p. 136).—*Caña Hueca* (Yacovleff and Herrera, 1934, p. 263).

Phragmites communis is found growing in moist locations in the coastal valleys where it often occurs in association with stands of *Gynerium sagittatum.* It also grows in the valleys with temperate climate in the mountainous areas of the interior. Its stems are commonly two to four meters in height with broad, flat leaves, and the large terminal plume-like panicles are usually purplish in color. Interesting features of the plant are its stolons and rhizomes, which often give rise to large colonies.

Two unworked stems of *Phragmites* have been identified from the site of Pachacamac (Towle, 1948). A large piece of matting from a grave at the Aspero site at Supe was made of the stems of this grass to which perpendicular strands of *Typha* were attached (Towle, 1954, p. 131). Yacovleff and Herrera (1934, pp. 263-264) report a Mochica vase with rhizomes, presumably those of this plant, painted upon the bowl. A container was found at Paracas Cavernas made of the stem of *Phragmites.* This contained needles or spines of different materials among which was *chonta, Guilielma ciliata* (*ibid.*, p. 311). They also describe cane tubes from the Chillón valley filled with powdered pigment of different colors and plugged with cotton (*ibid.*, p. 264). Likewise they mention combs and spindles made of these reeds recorded from Paracas (*ibid.*). In addition, these authors say that fragments of *Phragmites* have been recovered from Wayuri in the Department of Ica (*ibid.*).

SPOROBOLUS *R. Br.*
Vilfa Beauv.

Sporobolus is a genus consisting of about 100 species of annual and perennial grasses with narrow blades. These grasses are found in warm regions of both hemispheres. Six species are reported for Peru (Macbride, 1936, pt. 1, p. 173), several of which have been collected from coastal areas.

Sporobolus sp.

Wittmack (1880, Plate 106, fig. 14) identifies the grass found in a mummy bundle from Ancón as probably belonging to the genus *Vilfa,* now referred to as *Sporobolus.* This was used in large quantities to fill the spaces between the mummy and the wrappings. It is not possible to identify the species of this grass since his specimen as illustrated shows only the vegetative characters.

ZEA *L.*

Zea, a monotypic genus, indigenous to tropical America, is widely cultivated in both the Old and New Worlds.

Zea Mays *L.* Sp. Pl. (1753) 971.
Maize.—*Sara*, Quechua (Herrera, 1939, p. 313).

Maize, the largest of the cereals, is a coarse annual grass with solid, jointed stems and long, broad, two-ranked blades with wavy margins. The inflorescence is monoecious. The staminate flowers form large, spike-like panicles (tassels) at the top of the stem, while the pistillate flowers are borne in rows on a woody axis (cob) that develops in the leaf axil. This structure is surrounded by leafy bracts, which together form the husk, from the top of which emerges the thread-like styles (corn silk) of the enclosed flowers. The embryo and endosperm of the mature seed (grain) are surrounded by an outer covering (hull or pericarp), directly beneath which is a thin layer of protein (aleurone). Two kinds of endosperm are usually present: one hard and horny, the other soft and starchy in texture, the amount of each depending upon the kind of maize. There are several recognized types. These have been considered to be separate species of the genus or varieties of the single species, *Zea Mays*. They are pod corn, pop corn (either rice pop corn with pointed kernels or pearl pop corn with rounded, closely set kernels), flint corn, dent corn, soft or flour corn and sweet corn.

Maize was cultivated widely in the New World in prehistoric times, its area of distribution extending from the St. Lawrence and Missouri Rivers to the island of Chiloe in Chile. The plant is not known in the wild state. In fact, it has become so dependent upon man for its survival that it is unable to exist under natural conditions. Pod corn and pop corn are considered to be the most primitive types of the species. The characteristics that differentiate these forms from other varieties of maize are described as *tunicoid*, a term derived from the varietal name of pod corn, *Zea Mays* var. *tunicata*. Among these qualities are tapering ears with a heavy base, irregular rows of kernels, and prominent, membranous glumes. In addition, the stalk of the plant is often zigzag rather than straight and the joints may be enlarged. Sometimes the tassel is partially pistillate, and in some instances the ears show secondary branching.

Authorities on the genus (Mangelsdorf et al.) believe that the primitive characteristics exhibited by pod corn and pop corn are similar to those that the ancestral form of *Zea Mays* may possibly have possessed. Although no specimens of this ancestor of maize have been found, the plant is thought to have been originally widely dispersed throughout the temperate regions of South America, Central America and possibly Mexico, and at an early date was adopted by man as a crop plant. It has been postulated that either in one or more centers, before or after its domestication, this ancestral form crossed with species of a related genus, *Tripsacum*, producing hybrids that eventually developed into varieties of *Zea Mays*. The tripsacoid characteristics inherited by these hybrids were rounded, cylindrical ears with denser, more compact cobs, straight rows of kernels, shorter, stronger glumes, and an overall strengthening of the maize plant. Since there are a number of species of *Tripsacum* native to the supposed area of dispersal of ancestral maize, the hybridization between the two genera could well have occurred in more than one location. However, whether there were one or several centers of origin, the resulting hybrids were widely diffused over the region. These in turn were further differentiated by back-crossing or by additional strains of *Tripsacum*, resulting in the present-day varieties of *Zea Mays*.

After a detailed study of modern South American maize, Cutler (1946) distinguished seven races that are cultivated there at the present time. Several of these are grown in the Peruvian-Bolivian area. However, his study does not include the pop corns of which, as Cutler states, there are primitive as well as recent types. More recently, intensive investigations have been conducted by other botanists on all aspects of the maize problem in Latin America.

According to a recent report from the Escuela Nacional de Agricultura in Lima (Grobman et al. 1956), twenty-nine or possibly thirty races have been distinguished in Peru after a study of some 1,200 collections of Peruvian maize. These fall into two groups: highland races and lowland races. At high altitudes in the Andes, three or four ancient, indigenous races are still grown. These are all pop corns with marked primitive characteristics, and doubtless they were the progenitors of the other highland maize types recognized. The lowland races include those cultivated in the low coastal region and those grown in the lowlands east of the Andes as well.

The majority of the maize races described are indigenous, with little evidence of the intrusion of maize strains from other parts of Latin America. Certainly it would appear that maize has long been present in the Central Andean region of Peru, during which time there were developed, through independent evolution, the races found there today.

The history of this cultigen in Peru extends from prehistoric times. We have knowledge from the literature that maize was a crop plant in both the valleys of the Andes and the coastal region during the Conquest and the Colonial Period. Archaeological evidence has confirmed this as far as the coastal area is concerned. The desert conditions of the coastal strip have preserved large quantities of maize specimens, recovered as whole ears (sometimes enclosed in their husks), cobs, loose kernels, stalks, leaves and even whole or fragmentary tassels (pl. V). These give material for study in an attempt to reconstruct the type of maize that flourished in the area over a span of years.

Authors writing on the archaeology of the region, and more especially those that were concerned with a study of the plant remains recovered from the various sites, were quick to recognize the importance of maize in the economy of the people of prehistoric Peru. It was natural that they would attempt some form of classification by which the different types of maize could be described. Some of the earlier writers mention the similarity between the archaeological specimens and the maize grown in the Central Andean region at the present time. However, it was Wittmack (1880–87, 1888), in his study of the maize found at Ancón, who first developed a scheme for classification. This was based upon the morphological characters exhibited by the shape of the ears and characteristics of the kernels. He distinguishes three groups: (1) common maize, which he names *Zea Mays vulgata,* with kernels which are neither dented nor pointed although in some instances they are irregular in shape; (2) *Zea Mays peruviana,* which has short ears with pointed or beaked kernels; (3) and *Zea Mays umbilicata,* with seeds which have a groove in the outer surface. After classifying his material accordingly, Wittmack was left with a number of ears that possessed characters of two or possibly all three types; these are described as transitional forms. Rochebrune (1879), Costantin and Bois (1910), and Harms (1922) follow Wittmack's classification with some variations—or refinements—in an attempt to place the troublesome intermediate groups.

With the exception of those coastal sites attributed to the Epoch of Pre-Agriculture, there are few archaeological locations on the Peruvian coast that do not contain maize among their plant remains. The site of Huaca Prieta at El Brujo at the mouth of the Chicama River on the north coast is the outstanding exception. Here, Bird (Bennett and Bird, 1949, pp. 116–123) discovered a preceramic, pre-maize culture that was based upon a fishing, semi-agricultural economy. He gives the name of "Early Farmers" to the period it represented; its dates, determined by radiocarbon methods, are 2500–1200 B.C. (Bird correspondence, 1957). However, at a refuse heap near this mound at the same location, maize was discovered in Cupisnique levels dated at approximately 750 B.C. To the south, in the Virú Valley, maize specimens were recovered from each of four sites excavated by Strong and Evans (1952). These sites were Huaca Negra, which represents the Early and Middle Guañape Periods; a cemetery near Puerto Moorin, which has been described as essentially Salinar; Castillo de Tomaval, representing the Gallinazo Period; and Huaca de la Cruz, the lower levels of which are in the Gallinazo, the upper in the Mochica Period. The maize remains that were found consisted of cobs, tassels and corn husks. All of the specimens were recovered from Salinar, Gallinazo and Mochica levels. These included the cobs found at Huaca Negra, for although the site is designated as Guañape, the maize was from a grave of a later period. The prehistoric maize has small cobs, a low row count and usually possesses definite tunicate characters, a type often encountered in the archaeological sites on the Peruvian coast (Towle, 1952c, pp. 352–354).

On the central coast, maize specimens were recovered from the Aspero Midden and the Lighthouse sites at Supe, and these were found to belong to a single variety of the species (Towle, 1954, pp. 131–135). The maize found in the Cupisnique level at El Brujo and the specimens recovered from the Supe sites are quite similar, both representing a weak pod corn or a primitive pop corn, the type of primitive maize found in the earlier levels of other archaeological coastal sites. As mentioned above, Wittmack (1880–1887; 1888) describes in detail the maize remains recovered at Ancón by Reiss and Stübel and illustrates a tassel and various types of ears. Costantin and Bois (1910, pp. 252–256) examined eight ears from Pachacamac and describe them according to a modified Wittmack classification.

A large collection of maize specimens was collected from Inca and Inca-associated levels at Pachacamac and studied by Towle (1948). These specimens consist of ears, cobs, loose kernels, husks, tassels, leaves and stems, and present a range of types resembling the material found both in the earlier sites and that described from Ancón. An interesting specimen in this collection is a small, much-worn broom made of whole maize husks held together by a long leaf, probably that of *Typha angustifolia.* This was wrapped around the husks several times and then securely tied. The husks were remarkably tough and although the connecting tissue had disappeared through wear, the coarse strands remained. Further specimens from Pachacamac and from Chuquitanta are described by Harms (1922, pp. 162–163). These consist of ears, cobs, and loose kernels that were usually enclosed in small bags or gourd containers, often with other plant remains. Harms follows Wittmack's categories in describing this material. Maize remains described by Yacovleff and Herrera (1934, pp. 260–262) were recovered from the Chillón Valley. These authors remark that the specimens were markedly similar to the descriptions of the types examined by Wittmack from Ancón.

Large collections of maize have been recovered from sites on the southern Peruvian coast, but only a few writers on the archaeology of this area have given descriptions of the types found. An ear of maize from Wayuri, belonging to the Ica Period is described by Yacovleff and Herrera (1934, p. 262). Seven ears from Paracas Cavernas were examined by these authors (ibid., p. 260), as well as one from Paracas Necropolis (*ibid.*, p. 262). These specimens are of a flint corn with slightly dented kernels. Mangelsdorf (1942) examined ears from both Cavernas and Necropolis, and the shape of the ears as well as the characters of the kernels were similar to those studied by Yacovleff and Herrera. The three maize specimens found in a Paracas Necropolis mummy bundle examined by Towle (1952b, pp. 230–232) also have slightly dented kernels with a hard, flinty endosperm. The specimens from Paracas, like those mentioned above, represent a type of flint corn often found in pre-Columbian sites.

At Cahuachi in the Nazca Valley, maize was well represented throughout the explored area from a depth of 4.50 m. (Towle, 1956). Specimens of a similar type were included among the plant remains from Estaquería and Huaca del Loro, also located in the Nazca Valley and from Cacique, Pinilla and Ocucaje in the Ica Valley (*ibid.*).

From a study of the literature and actual specimens of prehistoric Peruvian maize, it appears that two major types or groups of this cultigen are represented in the coastal sites. The first type (Group A) is a weak pod corn or primitive pop corn. This maize is found in the earlier levels, and consists of small ears that have a tendency to be as thick as they are long. The diagonal rows on the slender cobs spiral either to the right or the left, and in some specimens they cross in two directions, the effect resembling a pine cone. Maize of this type has been recovered from the Ancón-Supe sites, from El Brujo and from the sites in the Virú Valley, all of which are mentioned above.

The second type (Group B) is found in the more recent archaeological levels. It resembles some of the forms of corn (excluding certain pop corns) that are grown in Peru today. The cylindrical ears are larger than those of Group A and they have a tendency toward straight rows. The larger grains, squared or oblong in outline, may have either a smooth or a dented outer surface. Still another type of kernel is beaked, and in some ears these beaked grains may be imbricated. There are many "intermediate" forms which possess grains with two or even three kinds of kernels. Wittmack's classification was based on Group B. Specimens of this group have been reported from Ancón, Pachacamac and other locations on the central as well as the southern coast. Anderson, in his study of the remains recovered from Arica on the north coast of Chile and from other sites on the south coast of Peru, also observed the existence of these two major groups (unpublished report). Cutler (1946, p. 282) does not differentiate between archaeological and modern maize. He assigns to his Altiplano race most of the archaeological Peruvian corn which has been recovered from sites within its area of distribution. According to Cutler, this race has a wide range of variation, certainly a characteristic common to prehistoric specimens.

The history and development of pre-Columbian Peruvian maize is still far from clear; the recognition of the two major groups only begins to clarify the situation. Consequently, the results of the current study of archaeological maize now being carried on by the Peruvian botanists are keenly awaited. Information so far obtained during the study of maize from Paracas and Nazca sites indicates that two or perhaps three races of pop corn, each with a distinctive pericarp color, were grown on the south coast in prehistoric times. Two of these types are now found only in the highlands (Grobman et al., 1956).

Of particular interest to archaeologists is the observation that Peru seems to be the home of pericarp color in maize (*ibid.*). In the earlier descriptions of specimens from archaeological sites, emphasis was often given to the variations in the color of the kernels. In fact, this characteristic was among those used when attempts were made to classify prehistoric maize.

Maize is the chief grain and food plant of Peru today. It is widely cultivated, and is grown not only in the highlands and inter-Andean valleys but as an important crop in the fertile valleys of the coastal region. Some of the common names used to designate various of the cultivated types often appear in descriptions in archaeological literature. Among the more commonly occurring are *hanka sara,* which is used for toasting, *Culli* maize, the kernels of which have a cherry-red pericarp and are used for making *chicha, morocho* maize grown for general use, and a sugar maize called *chuspillo.*

Our knowledge of the uses of maize in prehistoric Peru is fragmentary; the only information we have is supplied by the early writers of the Colonial Period. Wittmack (1888, pp. 332–333) gives a general summary of these references. It appears that bread made from maize was not ordinarily used but was reserved for special occasions. Safford (1917a, pp. 14–15) and Wittmack (1888, p. 329) mention maize cakes or bread found in prehistoric graves. Maize was eaten as a vegetable in both the milk stage and when mature, the full-grown ears being boiled (*mote*) or roasted (*camcha*). The stems are said to have been pressed for the sweet juice that they contained, and the mature kernels were used as a source of oil. Both a beer made from the fermented grain and an alcoholic beverage prepared from a mash of crushed kernels, chewed by the women, were consumed. This latter drink, known as *chicha,* is a popular beverage at the present time, and an excellent description of its preparation and use is given by Cutler and Cardenas (1947). Wittmack also says that roasted corn was used to make a beverage, but the method of its preparation is not known. He adds that the stalks and leaves of the corn plant were used as fodder.

It would be expected that a plant of the importance of maize would enter into the ceremonial and artistic life of the people, and such is the case. Both realistic and conventionalized representations of the corn plant and its various parts are elements of design on both pottery and textiles (Yacovleff and Herrera, 1934, pp. 258–259). These occur in high and low relief on pottery jars and are also molded in the round (pl. VII, A and B; pl. XV, A). There are illustrations of the use of the maize plant as a decorative motif in many archaeological descriptions of material that has been recovered. It is particularly common on jars of Mochica, Chimu, Nazca, Pacheco and Inca ware. Stalks of fruiting maize are represented on an Early Nazca textile (O'Neal and Whitaker, 1947, p. 320) and stalks of the plant were reproduced in gold by the Incas (Radin, 1942, p. 274). Herrera (1923, p. 445) reports identifying *sara* among the plants that were reproduced on *keros* found at Cuzco.

CYPERACEAE. Sedge Family.

Cyperus *L.*

Cyperus sp.

Bird (Whitaker and Bird, 1949, p. 3) mentions finding the small tubers of a species of *Cyperus* in the pre-ceramic levels at Huaca Prieta. These tubers, like the rhizomes of *Scirpus* and *Typha,* are used for food. Bird further reports that in the Chicama-Virú area the term "junco" is applied to a sedge identified as a species of *Cyperus* (Bird, correspondence, 1957).

SCIRPUS (*Tourn.*) *L.* Bulrush

Members of this genus are common plants of marshes and moist locations throughout the world. Macbride (1936, pt. 1, pp. 287–293) reports eighteen species occurring in Peru, several of these from the coastal valleys. The fibrous stems and leaves of certain bulrushes were used in prehistoric times for making rope, cordage, baskets, etc.

Scirpus tatora *Kunth* Enum. 2 (1837), 166.
Totora, Aymara.

The name *totora* refers in particular to *S. tatora* but also to other species of the genus (Herrera, 1939, p. 336; Macbride, 1936, pt. 1) and even to entirely different genera. Parodi (1932, p. 3) reports that in the Argentine the word "totora" stands for *Typha domingensis* (*T. angustifolia*). He further states that *S. tatora* is similar to those species referred to as "junco" in Buenos Aires. Bird (Correspondence, 1957) reports that rhizomes of a sedge found in the Chicama-Virú area are also known as "junco" (see *Cyperus*).

This plant is closely related to *Scirpus californicus,* a species found from Argentina to the United States. Some botanists, including Macbride (1936, pt. 1, p. 290), consider *S. tatora* a synonym of that species. However, Beetle (1945) and Parodi (1932) hold that the *totora* of Lake Titicaca is distinct. Its yellow-green culms, which are mostly sterile, often reach a height of from two to three meters and form dense stands about the shores of the lake. The pulpy stems of the bulrush are used in the construction of the famous reed boats or balsas made by the Aymara Indians of the southern part of Lake Titicaca and the neighboring Uru (LaBarre, 1948, p. 105).

Yacovleff and Herrera (1934, p. 294) state that the root-stocks of *S. riparius,* another related species if not a synonym, are used for food and the canes and leaves in various ways.

Scirpus sp.

A number of archaeological specimens made from the strong, fibrous culms and leaves of *Scirpus* have been recovered from sites on the coast of Peru. Two cords were found in Gallinazo levels at Huaca de la Cruz in the Virú Valley (Towle, 1952c, p. 354). Each specimen consisted of a number of culms tightly twisted together. In addition, a small piece of poorly preserved matting, probably made of the stems of *Scirpus,* was recovered from the Mochica level of the same site (*ibid.*). The fragile fiber matting from Burial 8 at the Aspero site at Supe was also probably constructed of the culms of a species of *Scirpus* (Towle, 1954, p. 137). Bands and ropes and even small bundles of unworked material, all consisting of stems and leaves of species of this genus, were found among the plant remains from the Inca and Inca-associated levels of Pachacamac (Towle, 1948). Yacovleff and Muelle (1932, pp. 133–134) point out that burials from both Paracas Cavernas and Paracas Necropolis were often provided with totora mats that vary in respect to technique of manufacture. A totora mat made of a simple twill weave accompanied the mummy bundle from Paracas Necropolis described by Towle (1952b, p. 232). In addition, two fans with handles made of the culms of this sedge were found in the bundle.

PALMAE. Palm Family.

EUTERPE *Gaertn.*

A genus consisting of 10 species of the tropical regions of Guiana, Brazil and the Andes.

Euterpe edulis *Mart.* Hist. Nat. Palm. 2 (1824) 33, t. 32.
Palmito (Yacovleff and Herrera, 1935, p. 51).

Euterpe edulis has a cylindrical, slightly thickened stem twenty to thirty meters in height. Its dark violet-colored fruits, known as *cocos de palmito*, are the source of an agreeable beverage. This species is found growing in the tropical areas of Brazil and the Guianas, and although it is not mentioned in the botanical literature as occurring in Peru, it might very possibly be found in the tropical forest areas of the Andean valleys or the eastern *montaña*. Yacovleff and Herrera (*ibid.*, fig. 52) reproduce a picture painted on an Inca wooden cup (*kero*), showing a hunting scene in a palm grove in which the hunters, armed with bows, are stalking monkeys and parrots that are eating the tender "hearts" at the top of the palm stems. The palm trees are painted in fair detail and these authors have identified them as *Euterpe edulis*.

GUILIELMA *Mart.*

The genus consists of probably three species of spiny feather-palms found in tropical America.

Guilielma ciliata (*Ruíz et Pav.*) *Wendl.* Kerch. Palm. (1878) 246.
Bactris ciliata (Ruíz et Pav.) Mart. Hist. Nat. Palm. 2 (1826) 95.
Chonta (Weberbauer, 1945, p. 147). This name also applies to *Guilielma* sp. (Herrera, 1939, p. 175), as well as to a number of other spiny palms (Dahlgren, 1936, p. 331).

This palm is characterized by its aerial roots and the numerous spines on the petioles of the leaves. It grows in the mountains of eastern Peru, and its dark, hard wood is used in various ways.

Rochebrune (1879, pp. 348, 354) describes work baskets from Ancón that contained spindles, the stocks of which were made of the wood of *Bactris ciliata.*

A container made of the stem of a reed, *Phragmites communis*, was found at Paracas Cavernas (Yacovleff and Herrera, 1934, p. 311). This contained needles or spines of different materials, among them *chonta* (*Bactris ciliata*).

Guilielma gasipaes (*HBK*) *Bailey*, Gentes Herb. 2 (1930) 187.
Guilielma speciosa Mart. Hist. Nat. Palm. 2 (1824) 82.
Peach Palm.–*Inchaui, inchauy* (Dahlgren, 1936, p. 347).–*Neije* (*ibid.*, p. 363). –*Pijuaio* (*ibid.*, p. 375).–*Pishoguayo*, Quechua for bird-fruit (Richard Spruce quoted by Cook, 1910, p. 309).

Guilielma gasipaes occurs in the West Indies, Central America and many parts of northern South America. This palm is one of the most useful plants of the Amazon Indians who have probably been responsible for its domestication throughout the Amazon region. In Peru it is found near the huts of the natives of the Loreto region, where it is grown for its edible fruits (Weberbauer, 1945, p. 147). Not only is the fleshy part of the fruit eaten but the hard, white kernel of the seed as well. The wood of this palm is dark brown in color and extremely hard (Popenoe and Jiménez, 1921). It was used by the Indians in pre-Columbian times to make spears and other weapons (Harms, 1922, pp. 164–65).

According to Rochebrune (1879, pp. 348, 354) the teeth of combs recovered from Ancón were made of the wood of *Guilielma speciosa* (*G. gasipaes*). The wood was also used, together with that of *Godoya obovata* and *Pineda incana*, for construction of the shafts of lances, clubs and other weapons. Saffray (1876, p. 401), in his description of the mummy of a warrior, states that the weapons that were found were made of the hard wood of the palm "macana." Furthermore, the combs that were recovered consisted of long spines or wooden teeth of "macana" held together with brown and white thread. Aside from the vernacular name "macana," there is no further clue as to what palm is being referred to. The word itself is derived from the native name for a sword, often edged with sharp teeth. In Venezuela it is applied to *Guilielma macana*, but the range of this palm does not extend into Peru. It would appear that the specimens described by Saffray were made of the dark wood of *Guilielma gasipaes*, although Harms (1922, p. 165) thinks that there should be more careful study of these archaeological specimens before one definitely accepts this identification.

Guilielma insignis *Mart.* Palmet, Orbign (1847) 71.
Bactris insignis (Mart.) Baillon Hist. Pl. 13 (1895) 305.
Chonta, Bolivia (Dahlgren, 1936, p. 331).–*Chonta de comer*, Bolivia (*ibid.*).– *Palma real, Bolivia* (*ibid.*).

This palm is found in Brazil and Bolivia. Levi-Strauss (1950, p. 472) reports that the natives of the upper Amazon and eastern Bolivia make their bows of its hard wood and that clubs and spears are often carved from it.

Harms (1922, p. 164) describes and illustrates (pl. 1, fig. 1) a specimen from Pachacamac made of the seeds of *G. insignis.* These are tied together with a string which is drawn through two eyes at the larger end of the seed; this object may have been used as a rattle or a toy.

PHYTELEPHAS *Ruíz et Pav.*

A genus of 3 species of northwestern South America.

Phytelephas macrocarpa *Ruíz et Pav.* Syst. Veg. (1798) 301.
Tagua palm, ivory-nut palm (Hill, 1952, p. 240).—*Palma de marfil* (Dahlgren, 1936, p. 231.)—*Humiro,* in central Peru (Weberbauer, 1945, p. 147).—Yarina, in northern Peru (*ibid.*).

The ivory-nut palm is found along the head waters of the Amazon in the tropical valleys of the *montaña* region of the Andes, at an altitude of between 1,800 and 2,000 meters. The fruit contains six to nine bony seeds which furnish the vegetable ivory used by both prehistoric and modern man. According to Rochebrune (1879, pp. 348, 354), specimens of whorls found among the artifacts from Ancón were identified as being made of the ivory-like seeds of this palm.

CYCLANTHACEAE. Cyclanthus Family.

CARLUDOVICA *Ruíz et Pav.*

About 40 species of these shrubby or herbaceous plants are found in tropical America.

Carludovica palmata *Ruíz et Pav.* Syst. Veg. (1798) 291.
Ludovia palmata Pers. Syn. Pl. 2 (1807) 576.
Bombonaje (Macbride, 1936, pt. 1, no. 3, p. 426).—*Appi-ttara* (*ibid.*).—*Bambonaje* (Herrera, 1939, p. 121).

This species is found growing in Peru in the inter-Andean regions and in the tropical forests of the eastern slopes of the Andes. It is palm-like in appearance and is often mistaken for a member of that family. Saffray (1876, p. 402) states that the small bags (*petacas*) included in a Peruvian mummy bundle were made of the petioles of this species. Rochebrune (1879, pp. 348, 354) says that the strands used in making work-baskets from Ancón were obtained from the leaf-stalks of *Ludovia palmata.*

ARACEAE. Arum Family.

XANTHOSOMA *Schott.*

This genus comprises about 38 tropical American species.

Xanthosoma sp.

Uncucha, a Peruvian name for *Xanthosoma* sp. (Weberbauer, 1945, p. 619). This name is used in the Argentine for *Xanthosoma sagittifolium* (Parodi, 1935, p. 131).

Both Weberbauer (1945, p. 619) and Herrera (1942a, p. 28) mention the growing of *uncucha* (*Xanthosoma* sp.) in the tropical regions of Peru. The leaves and also the roots are edible, and the latter, when dried, constitute one type of *chuño*. Safford (1917a, p. 24), in his discussion of the roots and tubers used in Peru in pre-Columbian times, describes a vase in the collection of the U.S. National Museum which is a representation of several tubers growing from a larger one. He says that these tubers have been doubtfully identified as *Xanthosoma*, but they may possibly represent the succulent roots of *Arracacia edulis*. Two specimens of tubers from a prehistoric site were identified by Costantin and Bois (1910, pp. 264–65; fig. 14, p. 264) as *Xanthosoma sagittifolium* Schott. It is unfortunate that the location of the site from which these specimens were obtained is not designated.

BROMELIACEAE. Pineapple Family.

ANANAS *Mill.*

A small genus of three species.

Ananas comosus (*L.*) *Merrill* Interpret. Rumph. Herb. Amboin. (1917) 133.
Ananassa sativa Lindl. Bot. Reg. 13 (1827) subpl. 1068.
Ananas sativus Schult. f. in R. and S. Syst. Veg. 7 (1830) 1283.
Pineapple.—*Piña*, Spanish (Yacovleff and Herrera, 1934, p. 267).—*Achupalla*, Quechua (Cook, 1925, p. 99).—*Ananas*, Brazil (Sauer, Carl O., 1950, p. 526).

The pineapple, endemic to South America, was cultivated by the natives of the tropics in pre-Columbian times. Its distribution on the Pacific Coast extended as far south as the lowlands of northern Peru. The first reference to the fruit was by Miguel de Estete in 1535 (1918; Yacovleff and Herrera, 1934, p. 268), who stated that " 'piñas,' though small," were to be found on the coast. No representations of this species have as yet been recovered from Peruvian archaeological sites. Wiener (1880, p. 601) states that fruits of the pineapple were used as models for pottery containers, but he does not support his remark either with descriptions or illustrations. Vargas, however (1947), reproduces an Inca design presumably part of the decoration of a *kero*, which he unquestionably identifies as "piña." In addition, he includes in an unpublished manuscript still another representation of the fruit of the pineapple which was painted on the side of a *kero* carved in the form of the head of a puma. Although in both instances the depiction of the *piña* is stylized, the characteristic leaves and fruit tend to confirm the identification.

TILLANDSIA *L.*

The genus *Tillandsia,* found in tropical and subtropical America, is composed of about 250 species most of which are epiphytic.

The stiff, fibrous stems and leaves of certain species of *Tillandsia* were used by the ancient Peruvians as a filling material for the false heads of mummy bundles, as well as for the cushions upon which the bundles were seated in the burial; and occasionally the loose plant material was used to form a layer upon which the mummy was placed (Safford, 1917a, pp. 13, 17). Representations of *Tillandsia,* and of *Opuntia,* a member of the Cactus family, are associated in war scenes on prehistoric jars (Harms, 1922, p. 165). Wittmack, in connection with his description of the specimens of *Tillandsia* found at Ancón (1880–87, pl. 106, figs. 15, 16), suggests that their presence in the mummy bundles points to the bundles having been brought from the highlands. However, a number of species of *Tillandsia* occur in Peru and several of these are native to the coast, where they are found growing in the sand and on rocky ledges. The distribution of these species seems to indicate that the plant material mentioned by Wittmack may have been of local origin.

Tillandsia Gilliesii *Baker* in Journ. Bot. 16 (1878) 240.
Tillandsia andicola Gill. *sensu* Wittmack, *non* Gill.

A pillow (24019, Peabody Museum, Harvard University) recovered from an ancient grave in Peru is filled with the leaves and stems of *T. Gilliesii.* A specimen referable to this species was found at Ancón in the wrapping material of a mummy bundle. This was identified by E. Moren as "probably *Tillandsia andicola* Gill." (Wittmack, 1880–87, pl. 106, fig. 15).

Tillandsia latifolia *Meyen* Reiss (1843) 437.

Two specimens of this plant, one an inflorescence, the other consisting of a number of leaves, were recovered from the Playa Grande level of the rubbish heap at Pachacamac (Towle, 1948). The presence of this species at the site has no particular significance except to indicate that the plant was available at this period. It may have had some definite use.

Tillandsia maculata *Ruíz et Pav.* Fl. Peruv. 3 (1802) 40, pl. 267.
Puca-huele (Herrera, 1939, p. 284).

Part of the inflorescence of a bromeliad, identified by E. Morren as "probably *Tillandsia maculata*" (Wittmack, 1880–87, pl. 106, fig. 16), was found within the wrappings of a mummy bundle from Ancón.

AMARYLLIDACEAE. Amaryllis Family.

AGAVE *L.*

There are about one hundred species of *Agave* found in the American tropics. These plants resemble in both habit and general morphological traits members of the related genus *Furcraea*. Yacovleff and Herrera (1934, p. 268) believe that early references to Peruvian maguey refer to *Furcraea* and not to *Agave*.

Agave americana *L.* Sp. Pl. (1753) 323.
Pacpa, used to designate both *Agave americana* and *Fourcroya andina* at Cuzco (Herrera, 1939, p. 262).

Apparently the only species of agave found in Peru is *A. americana*, which is considered to be an introduction from Mexico and Central America. In Peru the plant is usually found growing about human habitations where it is often planted as a hedge. Rochebrune (1879, pp. 348, 354) says that the fibers of plants of this species were used by the inhabitants of Ancón for making ropes, mats and baskets. According to Yacovleff and Herrera (see above), this refers to a species of *Furcraea*, probably *F. andina.*

FURCRAEA *Vent.*

This genus of fifteen species occurs in tropical America. *Furcraea* was named in honor of Count Antoine François de Fourcroy, a chemist at the Jardin du Roi in Paris. In the original publication, the name was spelled *Furcraea*, and, although it often appears as *Fourcroya,* the former spelling must be used in accordance with the International Rules of Botanical Nomenclature. Plants of this genus produce the stout leaf fibers that have made the group important economically in modern as well as in prehistoric times. In addition to their use for sandals and various types of cordage, Safford (1917a, p. 30) states that the ancient Peruvians made a cloth from the fibers of *Furcraea* which resembled the mummy cloth of the Egyptians. He also says that unworked, heckled fibers have been discovered in archaeological sites on the coast. Macbride (1936, pt. 1, no. 3, pp. 666–67) describes two species of this genus for Peru: *F. andina*, characterized by large, oblong-lanceolate leaves with prominent curved teeth along the edges, and *F. occidentalis,* with narrow oblong leaves, the edges of which are minutely prickled.

Furcraea andina *Trel.* in Bailey Stand. Cycl. Hort 3 (1915) 1305.
Chuchau (Macbride, 1936, pt. 1, no. 3, p. 666).–*Pacpa,* Cuzco (Herrera, 1939, p. 262).–*Chunta-pacpa,* Cuzco (*ibid.*, p. 176).

This plant held an important place in the economy of the ancient Peruvians. Not only were the fibers used to make sandals, twine, nets, etc., but the scape was employed in construction (Yacovleff and Herrera, 1934, p. 268). Sandals made of the fibers have been recovered from Cuzco (*ibid.*). A mummy from a site in the Chillón Valley had artificial hair made of the fibers of *F. andina;* these had been dyed blue with añil (*Indigofera* sp.) (*ibid.*). Both ropes and slings made from *Furcraea andina* are encountered in excavations on the coast (*ibid.*).

Furcraea occidentalis *Trel.* Bot. Jahrb. 50: Beibl. 3 (1913) 5.
Penca (Herrera, 1939, p. 274).

There appears to be no reference in the older literature to the use of this species of *Furcraea* in prehistoric times. However, of the two species described for Peru, it appears that this one is native to the hilly country of the western slopes of the Andes. Therefore it should be considered as possibly furnishing some of the fibrous material found in archaeological sites of the Pacific coast. Two fiber slings and a long plaited cord made of the fibers of *Furcraea* sp., probably *F. occidentalis,* were contained in a mummy bundle from Paracas Necropolis (Towle, 1952b, p. 233).

Furcraea gigantea *Vent.* in Bull. Soc. Philom. 3 (1793) 65.
Furcraea foetida (L.) Haw. Syn. Pl. Succ. (1812) 73.
Piteira, Brazil and other Portuguese-speaking countries (Dewey, 1943, p. 28).—
Piteira gigante (*ibid.*).

This species of *Furcraea* is native to the American tropics, particularly to Brazil.

In spite of references to Peruvian archaeological specimens of the fibers of plants of this species, neither Macbride (1936, pt. 1, p. 223) nor Weberbauer (1945) mention it in their accounts of the flora of Peru. Saffray, for example (1876, p. 402), states that two "alpargatas" or sandals made of fibers of *Fourcroya foetida* were included among the specimens which he examined. Rochebrune (1879, p. 348) mentions *Fourcroya foetida* Haw. as a source of industrial fibers, and also says (*ibid.*, p. 354) that this species furnished the ligneous fibers from which sandals were made.

CANNACEAE. Canna Family.

CANNA *L.*

Plants of this genus, comprising about thirty species, are native to tropical America where they are cultivated as ornamental and as food plants. In Peru, the canna is found on the coast and in the temperate valleys of the Andes to an altitude of 2,000 meters. Nine species are described by Macbride (1936, pt. 1, no. 3, p. 738). Two of these, *C. edulis* and *C. indica,* both known as *achira,* are closely related, and are cultivated to a limited extent for their edible tubers. When boiled, they have a sweetish taste, but because of their fibrous character, they are inferior to the sweet potato. Notwithstanding, these starchy tubers have been a food item in both modern and pre-Columbian Peru (pl. I, 6; pl. X, A).

Canna edulis *Ker-Gawl* Bot. Reg. 9 (1823) 64, pl. 775.
Achira, a name given to those species of canna that are cultivated for food, especially *C. edulis* (Macbride, *ibid.*).

Rochebrune (1879, p. 348) mentions *Canna edulis* as one of the food plants of the ancient Peruvians and adds (p. 352) that tubers of this species are found in archaeological sites. Two vases representing such tubers are described and illustrated by Safford (1917a, p. 14, and pl. III; p. 24 and fig. 13). These specimens were recovered from the coastal area in the vicinity of Trujillo and Chimbote.

Canna indica *L.* Sp. Pl. 1 (1753) 59.
Achira (Yacovleff and Herrera, 1934, p. 311).

Macbride (1936, pt. 1, no. 3, p. 750) states that this species is found in Peru only in cultivation. The edible tubers of this plant are represented in the ancient pottery of Nazca and Chimu (Yacovleff and Herrera, 1934, pp. 311–12). These vases are either in plain globular shapes or have been molded to represent one or more tubers. The surfaces of both types are usually decorated with designs inspired by the lines and ridges that mark the epidermis of the tuber. Still other pictorial representations of a similar nature have been found (*ibid.*).

Canna sp.

It is often impossible to differentiate between these two species of *Canna* as far as archaeological material is concerned.

A number of tubers of the canna were recovered from Inca and Inca-associated levels of the refuse heap at Pachacamac (Towle, 1948). Bird found a rope made of a twisted canna leaf among the specimens of cordage from this site (Bird in conversation, 1948). He also found in the pre-ceramic levels at Huaca Prieta leaves and fragments of the leafy covering of the rhizomes of the canna (Whitaker and Bird, 1949, p. 3). A small bundle of folded petioles of *achira* was among the specimens recovered from the Maranga level at Vista Alegre on the central coast, while part of a rhizome was included with the Decadent Maranga materials from this site (Towle, 1958). Tubers of *achira* have been recovered from the sites of Cahuachi, Estaquería and Huaca del Loro in the Nazca region of the south coast (Towle, 1956).

PIPERACEAE. Pepper Family.

Piper *L.*

This large genus occurs in the warmer parts of both hemispheres. However, few species are of economic importance.

Piper asperifolium *Ruíz et Pav.* Fl. Peruv. 1 (1798) 37, pl. 56.

This plant is said to have been used medicinally in pre-Columbian Peru (Rochebrune, 1879, pp. 347, 351, 352).

Piper lineatum *Ruíz et Pav.* Fl. Peruv. 1 (1798) 35, pl. 60.
Schilleria lineata Kunth, Linnaea 13 (1839) 704.
Cordoncillo, common name for this and many other species of the genus (Herrera, 1939, pp. 153–56).

Rochebrune (1879, pp. 347, 355) states that the leaves of this species, when combined with the blue and yellow dyes obtained from other native plants (*Dicliptera Hookeriana, D. peruviana* and *Lafoensis acuminata*), furnished the ancient Peruvians with a green dye.

SALICACEAE. Willow Family.

The willows comprise about 170 species of the temperate zones of both hemispheres.

Salix *L.*

Salix chilensis *Mol.* Sagg. Chil. (1782) 169.
Salix Humboldtiana Willd. Sp. Pl. 4 (1805) 657.
Sauce (Macbride, 1936, pt. 2, no. 2, p. 261).

Salix chilensis is one of the two native willows of Peru (*ibid.*). Both wild and cultivated trees are found to an altitude of 3,000 meters. Yacovleff and Herrera (1934, pp. 271–72, fig. 10) mention the discovery of osier baskets in the cemeteries of Paracas. These have a framework of willow stems (*S. Humboldtiana*) averaging two to three mm. in diameter and interlaced with the culms of *totora.*

PROTEACEAE. Protea Family.

Roupala *Aubl.*

Nine species of this genus are described for Peru (Macbride, 1936, pt. 2, no. 2, pp. 371–75). Some of these stout shrubs grow to an altitude of 1,400 meters on the western slopes of the Andes, while others are found in the eastern and northern sections up to an altitude of 2,000 meters (Weberbauer, 1945, p. 159).

Roupala ferruginea *HBK.* Nov. Gen. et Sp. 2 (1817) 152.

Rochebrune (1879, pp. 347, 355) reports that a specimen of the bark of this plant was found at Ancón. He adds that this was a source of a brownish dye in pre-Columbian Peru and that it is used for the same purpose at the present time (Saffray, 1876, p. 407).

CHENOPODIACEAE. Goosefoot Family.

Chenopodium *L.*

This genus of about sixty species of both hemispheres includes several that are native to Peru. The most important economically are *C. Quinoa*, commonly known as "quinoa," and *C. pallidicaule,* known as "cañihua" or "cañahua." Although much less valuable than *quinoa,* the seeds of *cañihua* are also frequently used for food. The plant is smaller than *quinoa* and produces its seeds in loose panicles. It is largely grown in temperate areas, while *quinoa* is found in the colder altitudes of the Andes of southern Peru and Bolivia. Although *cañihua* has not been reported from archaeological locations, it was in all probability a source of food in the Andean highlands in prehistoric times.

Chenopodium Quinoa *Willd.* Sp. Pl. 1 (1797) 1301.
Quinoa, quinua (Herrera, 1939, p. 294).

The nutritious seeds of this plant furnish a staple food for a large segment of the native population of Peru, replacing maize in the higher altitudes of the Andes. The red, white or black seeds are used whole to thicken soup, to make *chicha,* or to be ground into flour. The ashes of the stalks may be combined with the leaves of coca, a combination that is said to increase the flavor of the latter.

Quinoa was used in prehistoric times in Andean South America, and specimens of the plant have been recovered in several archaeological sites in Peru, Chile and Argentina (Hunziker, 1943). Safford (1917a, p. 15) reports that in 1887 terminal clusters of *quinoa* were found in a prehistoric burial at Arica, Chile; these were in association with specimens of llama wool and other objects characteristic of the Peruvian plateau area. More recently, Bird reports the occurrence of *quinoa* seeds in fabrics from Arica: one from a disturbed burial intrusive in Arica II at Playa Miller and the other from the "Black Refuse" level at Pichalo (Bird, Correspondence, 1957). Seeds, leaves and stems of this plant were included in the mummy bundles recovered at Ancón (Wittmack, 1880–1887, pl. 106, fig. 20). Rochebrune (1879, pp. 347, 352) further states that small cakes made of a combination of the flours of *quinoa* and maize were found at this site. He adds that small bags containing leaves of coca (*Erythroxylon Coca*), lime and quinoa flour were also recovered. At the site of Chiripa in the South Highlands, a bin containing *quinoa* seeds was discovered within one of the walls of a house (Bennett and Bird, 1949, p. 142; Towle, 1957).

AMARANTACEAE. Amaranth Family.

Amaranthus *L.*

The species of *Amaranthus*, of which there are about sixty, are herbaceous plants of the temperate and tropical regions of both the Old and New Worlds.

A few species have been domesticated. The variability existing among these cultivated forms has resulted in considerable confusion and a lack of clear, specific definition.

Amaranthus caudatus *L.* Sp. Pl. (1753) 990.
Bledos (Cobo, 1890, Lib. IV Cap. II, p. 337).—*Inca-pachaqui*, Bolivia (Ames, 1939, p. 34).—*Quihuicha*, Cuzco (Weberbauer, 1945, p. 619).

This herbaceous species is grown as a seed crop in the Andean region of Peru, Bolivia and northwestern Argentina. The plants are characterized by the long, drooping, tail-like inflorescences from which the species derives its name. The seeds are predominantly pale, but may be red or dark-colored (Sauer, Jonathan D. 1950, p. 603). Cook (1925, p. 36) states that the seeds of *A. caudatus* will pop like kernels of maize and that their taste resembles that of this latter cereal.

Although archaeological specimens of *Amaranthus* have not been recovered in Peru, this plant was possibly a grain crop of some importance in the temperate valleys of the highlands in prehistoric times (Herrera, 1940, p. 229). The one reference to the grain amaranths in the writings of the Colonial Period occurs in Cobo (l.c.). He makes the distinction between *quinoa* and *bledos*, the descriptive word for amaranth used by the early chroniclers of Mexico. Cobo mentions that the *bledos* of Peru had red and white seeds that were a common food of the Indians.

Amaranthus cruentus *L.* Syst. Nat. ed. 10, 2 (1759) 1269.
Amaranthus paniculata L. Sp. Pl. ed. 2 (1763) 1406.
Bledos, Spanish (Yacovleff and Herrera, 1935, pp. 84–5).—*Jattacco* (*ibid.*)

The American distribution of this species is given as Central Mexico to Panama and in tropical and subtropical South America (Standley, 1917). Yacovleff and Herrera (*ibid.*) identify the "bledos" referred to by Cobo (1890, Cap. II, p. 337) as *Amaranthus paniculatus.*

BASELLACEAE. Basella Family.

Ullucus *Lozano*

A monotypic genus.

Ullucus tuberosus *Lozano in* Caldas, Seminario de Nueva Granada (1809) 185.
Lisas, papaslisas, Southern Peru and Bolivia (Hodge, 1946, p. 218).—*Ullucu*, Quechua (*ibid.*).—*Olluco*, Cuzco (*ibid.*).—*Melloco*, Ecuador and neighboring regions of Colombia (*ibid.*). In addition to these general names for the plant, Herrera (1921, pp. 70–72), in his description of the flora of Cuzco, records a number of other native names for various cultivated and wild forms.

The tubers of this plant, endemic to the Andes, constitute one of the staple foods of the natives of that region. Although it is cultivated widely at high altitudes from Bolivia and Central Peru to Ecuador and Colombia, it does not surpass the potato as a root crop. The tubers vary in both shape and color. They may be either small and round like potatoes or elongated and curved. The predominating colors are either pale magenta or yellow, the latter sometimes occurring with magenta spots. The tubers of *Ullucus* are prepared and eaten like potatoes and, like the latter, are often made into *chuño*. The fact that this plant was used as a source of food in pre-Columbian times is attested by the tubers of *Ullucus tuberosus* found in archaeological sites. According to Rochebrune (1879, pp. 347, 352), several of these were recovered at Ancón. Harms (1922, p. 167) describes in detail the tubers that he found at Chuquitanta, a site on an estuary of the Río Corabaillo in the vicinity of Lima, illustrating two of these on plate 1, figure 2. It is safe to assume that all of these specimens had been brought from the highlands in the form of *chuño*. Yacovleff and Herrera (1935, p. 47) do not mention finding archaeological material of *Ullucus*, but they do describe and reproduce (1934, fig. 28, b) a coastal Tiahuanaco painting from Pacheco in which the design consists of stems, leaves and tubers, both round and oblong; the authors identify the plant depicted as *Ullucus*. This identification is questioned by Hodge (1946, p. 223) on the grounds that the leaves are ovate rather than heart-shaped as in the case of *Ullucus;* he believes that the representation is more likely a potato.

ANNONACEAE. Custard-Apple Family

Annona *L.*

Annona is a large genus comprising about sixty species mostly of tropical America. The term "custard-apple" has been given to several of these because of their agreeable, soft, custard-like fruit. The most commonly-known species are *Annona Cherimolia*, the cherimoya; *A. muricata*, the soursop; *A. squamosa*, the sweetsop; and *A. reticulata*, the custard-apple or bullock's heart. Although there are references to all four of these in the modern botanical literature of Peru, only the first three appear to have been used in prehistoric times.

Annona Cherimolia *Mill.* Gard. Dict. ed. 8. (1768) no. 5.
Cherimoya; Chirimoya (Herrera, 1939, p. 174).

The cherimoya, a small tree bearing edible, heart-shaped fruits, occurs naturally in the temperate Andean valleys of Ecuador and Peru. A number of varieties have been developed. Safford (1917a, p. 19) describes three different kinds of cherimoyas from the United States National Museum specimens that were recovered from graves at Ancón. Five seeds of cherimoyas are described by Costantin and Bois (1910, p. 257) from the site of La Rinconada near Lima; one of these is illustrated by the authors (*ibid.*, fig. 10). Safford (*ibid.*, p. 19) also mentions a prehistoric funeral vase from Peru which he thinks was cast from the fruit of

Annona Cherimolia. The cherimoya is included among the fruits given by Wiener (1880, p. 601) as having been used as models for clay jars in prehistoric Peru.

Annona muricata *L.* Sp. Pl. (1753) 536.
Soursop (Hill, 1952, p. 418).—*Guanabana* (Herrera, 1939, p. 188).

The soursop is widely cultivated from Central America to the coastal valleys of southern Peru for its edible fruits, which are large, ovoid and spiny. These have often been reproduced in the black pottery from Chimu on the northern coast of Peru (Yacovleff and Herrera, 1934, pp. 275–76; fig. 1, f). Safford refers (1917a, p. 19) to two funeral vases, one of which closely resembles *A. muricata.*

Annona squamosa *L.* Sp. Pl. (1753) 537.
Sweetsop or sugar apple (Hill, 1952, p. 417).

The sweetsop, a native of tropical South America and the West Indies, is valued highly for its delicious fruit. Although apparently no archaeological specimens have been actually identified as belonging to this species, Wiener (1880, p. 601) states that the fruit of *A. squamosa* was used as models for pottery jars in prehistoric times.

LAURACEAE. Laurel Family.

Aniba *Aubl.*

The genus *Aniba* comprises about 60 tropical American species.

Aniba Puchury-minor *(Mart.)Mez* in Jahrb. Bot. Gart. Berl. 5 (1889) 70.
Nectandra Puchury-minor Nees et Mart. ex Nees, Syst. Laur. (1836) 336.
Acrodiclidium Puchury-minor sensu Safford.
Moena amarilla, Peru (according to Williams as stated by Kostermans, 1938, p. 904). This name, according to Herrera (1939, p. 240), is also applied in Peru to certain members of the genus *Nectandra.*—*Puchery,* Brazil (Kostermans, 1938, p. 904).

This species is found in the upper Amazon basis and its tributaries. The seeds, which are sold as Pichurim beans, are used in medicines for diarrhoea and dysentery.

Costantin and Bois (1910, pp. 258–259; fig. 11) describe three pierced cotyledons of seeds which they identify as those of *Nectandra Puchury-minor* Nees et Mart. Harms (1922, p. 168), however, after citing this reference, adds that the specimens probably should be identified as cotyledons of *Aniba Puchury-minor* Mez. Safford (1917a, p. 22; fig. 8) says that in the collections at the American Museum of Natural History and the United States National Museum there are strings of bean-like cotyledons of seeds of a lauraceous tree closely related to *Acrodiclidium Puchury-minor* and more often referred to the genus *Nectandra.*

NECTANDRA *Rottb.*

A genus of about 60 species of tropical and subtropical America.

Nectandra Pichurim (*HBK*) *Mez,* in Jahrb. Bot. Gart. Berl. 5 (1889) 449.
Isula-micuna (Herrera, 1939, p. 208).—*Pishco-nahui-muina* (*ibid.,* p. 281).

A pottery jar from Moche is illustrated by Yacovleff and Herrera (1935, p. 89); they identify the bean-like clay forms that decorate the circumference of the vessel as the cotyledons of *N. Pichurim.* These are depicted as strung on a cord and resemble the necklaces of similar specimens that have been reported from other archaeological sites in Peru.

Nectandra reticulata (*Ruíz et Pav.*) *Mez,* in Jahrb. Bot. Gart. Berlin 5 (1889) 404.
Nectandra mollis Nees Syst. Laur. (1836) 287.
Huarme tashango (Herrera, 1939, p. 198).—*Muena,* Loreto (*ibid.,* p. 245). This name also is applied to certain other species of *Nectandra* and of *Ocotea* (*ibid.*).

Wittmack (1888, p. 327) reports four pierced cotyledons among the specimens in the Acland collection from Ancón, discovered with the mummy of a child. He identifies these as perhaps *Nectandra mollis* Nees.

Nectandra sp.

A necklace of nine pairs of pierced cotyledons found at Pachacamac and a similar necklace recovered at the site of Chuquitanta are described by Harms (1922, p. 168). However, he does not attempt to identify the species of *Nectandra* to which these specimens belong. Two pierced cotyledons of a *Nectandra* seed strung on a cotton string (pl. IV, 8) were found in a work basket in a cemetery located at the Hacienda Grana in the Chancay Valley. This specimen is at the Botanical Museum, Harvard University.

PERSEA *Mill.*

A genus of approximately 10 species of the American tropics.

Persea americana *Mill.* Gard. Dict. ed. 8 (1768).
Persea gratissima Gaertn. Fruct. 3 (1807) 222.
Avocado.—*Aguacate* (Spanish).— *Palto, palta, paltai* (Macbride, 1938, vol. 13, pt. 2, no. 3, p. 875).—*Huira palta* (*ibid.*).

The avocado, a native of tropical America, is often cultivated for its edible brownish-green, pear-shaped fruit, which has a single large seed surrounded by a soft, perishable pulp. This fruit was much used for food in prehistoric times, and remains of the seeds, either whole or broken into their two convex cotyledons, are frequently found in archaeological sites on the coast of Peru. Rochebrune (1879, pp. 347, 351) identifies whole seeds and separated cotyledons among the plant remains from Ancón as *P. gratissima* var. *oblonga* Gaertn. Cotyledons of the avocado were also found among the specimens from a Gallinazo site (Castillo de Tomaval) in the Virú Valley (Towle, 1952c, p. 355) (pl. III, 2), while a seed of this fruit was recovered from the refuse of a prehistoric cemetery in the Chillón Valley (Yacovleff and Herrera, 1934, p. 276). Harms (1922, p. 168) identifies avocado leaves among those found in a mummy bundle from Pachacamac and mentions a bag from the same site that contained leaf fragments of this plant. Additional evidence of the presence of the avocado at Pachacamac is provided by several whole seeds and cotyledons found in the Inca and Inca-associated levels (Towle, 1948). Leaves of the avocado were used to fill the false head of a mummy from Zapollango on the central coast (Peabody Museum, Harvard University). Bird discovered the avocado in Cupinsnique levels at Huaca Prieta (Whitaker and Bird, 1949, p. 6). Specimens of both whole and broken cotyledons were among the remains recovered from Cahuachi in the Nazca Valley (Towle, 1956). According to Wiener (1880, p. 601), the ancient Peruvians molded clay jars to resemble this fruit, but Yacovleff and Herrera (1934, p. 277) state that they know of no artistic representation of the avocado from prehistoric Peru.

ROSACEAE. Rose Family.

KAGENECKIA *Ruíz et Pav.*

Kageneckia is a genus of three species native to Chile. *K. lanceolata* is also found in the western and inter-Andean sections of Central Peru.

Kageneckia lanceolata *Ruíz et Pav.* Syst. Veg. (1798) 290.
Lloque and its variants (Yacovleff and Herrera, 1935, p. 37).—*Uritumicuma* (Herrera, 1939, p. 347).

This shrub has a flexible stem and elongated branches which bear many short, thick-leaved, flowering branchlets. The small flowers, in loose clusters, develop solitary fruits covered with short, dense, woolly hairs. Walking sticks and clubs are made from the tough wood of this plant and it is also employed in the construction of hanging bridges. An infusion of the bitter bark and leaves is a remedy for fever. The leaves furnish a black dye. *Macanas* (staffs or clubs), made of this wood and set with stone, are reported by Yacovleff and Herrera (*ibid.*) as having been frequently recovered from coastal Peruvian graves. These authors make the interesting observation that one of the Inca Emperors was named *Lloque yupanqui*, the first part of the name signifying strength.

LEGUMINOSAE. Pulse Family.

ACACIA *Willd.*

A cosmopolitan genus of about 500 species.

Acacia macracantha *Humb. and Bonpl. ex Willd.* Sp. Pl. 4 (1806) 1081.
Espino, Dept. of Ica (Weberbauer, 1945, p. 168).—*Faique*, northern coast of Peru (*ibid.*).—*Huarango*, Central Peru (*ibid.*).

This spiny tree with pinnate leaves and small yellow flowers is widely distributed on the western slopes and in the inter-Andean valleys of Peru as well as along the rivers of the coast. Wittmack (1880–1887, pl. 107, fig. 1, 2, 3) illustrates pods and seeds of a leguminous plant recovered at Ancón; he does not attempt to identify these beyond saying that they are of "a Mimosaceae." Harms (1922, pp. 170–71), however, compares these specimens with modern plant material of *Acacia macracantha*, and although he does not identify them with absolute certainty, he thinks that in all probability the archaeological material belongs to this species.

ANDIRA *Lam.*

A genus comprising about twenty-five tropical American species.

Andira inermis (*Wright*) *HBK.* Nov. Gen. et Sp. 6 (1824) 385.
Andira stipulacea sensu Rochebrune and Wittmack.
Quinillo colorado, Peru (Record and Hess, 1943, p. 232).

Rochebrune (1879, pp. 346, 352) mentions *Andira stipulacea* among the medicinal plants used by the ancient Peruvians. Wittmack (1888, p. 346) repeats this reference and adds that the plant was used as a vermifuge. It appears that there is some confusion in the identification made by Rochebrune, since *Andira stipulacea* is a species found in eastern Brazil and is not included in the botanical literature for Peru. What he may have had in mind was another species of the genus *Andira inermis*, a tree widely distributed in the mountainous areas of Central America and northern South America, including northern Peru. This plant possesses the same medicinal qualities as *A. stipulacea*, the seeds and bark being used as an emetic and vermifuge.

ARACHIS *L.*

A genus of about ten species found in South America.

Arachis hypogaea *L.* Sp. Pl. (1753) 741.
Peanut.—*Cho'kopa*, Aymara (Mejía Xesspe, 1931, p. 11).—*Inchis*, Quechua (*ibid.*).—*Mani*, Spanish. This name was adopted by the Spanish explorers from the tribes of the West Indies.

This cultigen of tropical and subtropical America is usually thought to have originated in Brazil, where several related wild species are found today. The plant is now cultivated as an annual food and forage crop in many parts of the world, with more than twenty varieties being recognized. Carl O. Sauer (1950, p. 499) considers that the peanut, in its New World distribution, may possibly have been part of the manioc (bitter and sweet) cultivation complex. From archaeological evidence, the peanut appears to have been widely used on the coast of Peru in pre-Columbian times. Remains of the pod are among the most commonly encountered plant remains, and in addition, we find pottery vessels decorated with representations of the pods (pl. VIII, B; pl. IX, A) as well as textile designs that depict parts of the peanut plant.

The usual variety of *Arachis hypogaea* found in the coastal sites has a long, slender, reticulated pod with either one or two hump-like protuberances on the dorsal surface (pl. II, 1). Apparently this is similar to a type found in the Orient today (Ames, 1939, pp. 46–48). However, a smaller variety is also reported, in which the shell is only slightly reticulated and lacks the dorsal humps. Specimens of this latter type, which resemble certain varieties now cultivated in North America, were recovered from a site at Supe that belongs to the Early Ancón Period (Towle, 1954, p. 135). Numerous specimens of the peanut have been recovered from Ancón (Rochebrune, 1879, pp. 346, 350; Wittmack, 1880–1887, pl. 107, fig. 4, 5; Wittmack, 1888, pp. 338–39; Safford, 1917a, p. 16) and Pachacamac (Harshberger, 1898, p. 3). Some specimens have been definitely correlated with the Inca and Inca-associated levels of the latter site (Towle, 1948) and peanuts are reported from Cupisnique levels at Huaca Prieta (Whitaker and Bird, 1949, p. 6). A broken pod was found in a grave at the Gallinazo site of Castillo de Tomaval in the Virú Valley (Towle, 1952c, p. 355). Yacovleff and Herrera (1934, p. 280) state that well-preserved peanuts have been found among the plant remains in the graves of the Chillón Valley and at Paracas, while Towle (1952b, p. 235) describes peanuts that were found in a mummy bundle from Paracas Necropolis. Saffray (1876, p. 402) remarks that pods of this legume are often contained in small string bags with other plant remains. Harms' description (1922, p. 172) of a small gourd containing three pods and a net bag filled with remains of food plants including peanuts illustrates this point; both of these specimens were recovered from Chuquitanta. Harms (*ibid.*, p. 173) also describes two earthenware jars, one from Trujillo, that are decorated with extraordinarily life-like reproductions of peanut pods. Safford (1917a, pp. 16–17) refers to earthenware pans, one with painted peanuts on the handle, which were recovered from a grave at Chimbote; he believes that these were used for parching peanuts. Actual remains of whole and broken pods of *mani* were found in the general excavations at Cahuachi and Huaca del Loro in the Nazca Valley and also in burials at the former site and at Ocucaje in the neighboring Ica region (Towle, 1956). The plant and pods of *A. hypogaea* were depicted on Early Nazca textiles (O'Neale and Whitaker, 1947, pp. 311–20).

CAESALPINIA *L.*

Caesalpinia, a genus of more than one hundred species, occurs in the tropics and subtropics of both hemispheres.

Caesalpinia Paipai *Ruíz et Pav.* Fl. Peruv. 4 (1802) pl. 375.
Caesalpinia corymbosa Benth. Pl. Hartw. (1843) 117.
Charan, in Piura and Tumbes (Weberbauer, 1945, p. 170).—*Pai-pai,* in Libertad (*ibid.*).

Caesalpinia Paipai occurs as shrubs or small trees along the rivers of the northern coast of Peru. The slender pods are short stiped, seven cm. or longer, rather fleshy and somewhat constricted. The wood is used in carpentry and the pods furnish a black dye. Several pods referred to *Caesalpinia corymbosa* have been recovered from archaeological sites. A pod found at Ancón is identified by Harms (1922, p. 171) as belonging to this species. Similarly, he identifies a short, thick pod from Pachacamac that had been placed in a gourd container with specimens of peanuts and beans. In addition, he describes several pods from Chuquitanta that were included in a gourd shell, and also four others that were in a small net bag.

Caesalpinia spinosa (*Mol.*) *Kuntze* Rev. Gen. 3, pt. 2 (1898) 54.
Coulteria tinctoria HBK. Nov. Gen. et Sp. 6 (1824) 261, p. 569.
Caesalpinia tinctoria (HBK.) Dombey ex DC. in syn., Prodr. 2 (1825) 481.
Tanino (Macbride, 1943, vol. 13, pt. 3, no. 1, p. 196).—*Tara,* Central and south coasts of Peru (Weberbauer, 1945, p. 169).—*Taya, North coast* (ibid.).

Caesalpinia spinosa, a stocky shrub or small tree, is frequently found growing along the river banks of the *lomas* region of the Peruvian coast. It is also commonly planted. This species has a spiny, gray bark, densely leafy branchlets and slender racemes of reddish-yellow flowers. The red or reddish pods are used for tanning and as a source of dye; the powder found in the pods is said to be used as an eye wash (Macbride, 1943, vol. 13, pt. 3, no. 1, p. 196). Rochebrune (1879, pp. 346, 355) reports finding the pods of *Coulteria tinctoria* at Ancón. According to Wittmack (1888, p. 347), these pods were used as a source of a black dye. Harms (1922, p. 171) says that gourd shells were found at Chuquitanta that contained leaves of *Caesalpinia tinctoria.*

CANAVALIA *Adans.*

Members of this genus of approximately twelve species are bushy or climbing herbs found in the tropical and subtropical regions of the Old and New Worlds. The valves of the pods are characterized by one to four longitudinal ridges, which often parallel or even unite with one or both of the sutures. The globose or ellipsoidal seeds are sometimes separated by membraneous tissue within the pods. The seeds of the different species vary in color.

Canavalia sp.

Pallar de gentiles, the vernacular name given for *Canavalia* sp. (Yacovleff and Herrera, 1934, p. 290).

Canavalia beans have been identified among the plant remains from many of the archaeological sites on the Peruvian coast (pl. II, 6). These specimens, with either a black or a brown hilum, may be yellow (with or without brown spots), reddish brown, brown, dark gray or sometimes black. The beans are found as dissociated specimens or with other plant remains in gourd containers or in net bags. Costantin and Bois (1910, pp. 250–51; fig. 3) describe two seeds, one black, the other a yellow-brown, as *Canavalia ensiformis,* the jack-bean. Other *Canavalia* seeds, yellow with brown spots or gray, have also been assigned to this species. These were recovered from Paracas and Zappalli, a site in the Chillón Valley (Yacovleff and Herrera, 1934, pp. 290–91). Similar specimens from Chuquitanta and Paracas are described by Harms (1922, pp. 173–74). However, he dismisses the possibility of their being the jack-bean, since the seeds of this species are white with a black hilum. In lieu of comparative plant material from Peru, he notes a similarity in color between his specimens and the seeds of *Canavalia obtusifolia,* a species native to Asia. Brown, yellow and mottled seeds of *Canavalia* as well as a complete pod were recovered from Inca and Inca-associated levels of Pachacamac; these were identified as *Canavalia* sp. (Towle, 1948). Other specimens, similarly identified, were found in Gallinazo and Mochica levels at a site (Huaca de la Cruz) in the Virú Valley (Towle, 1952c, p. 355). In the Nazca region of the south coast, beans of *Canavalia* were commonly found in the general excavations at Cahuachi and at Huaca del Loro and Estaquería (Towle, 1956). Bird reports the finding of at least three varieties of beans ("all *Canavalia*?") in the pre-ceramic levels at Huaca Prieta (Whitaker and Bird, 1949, p. 3). Lawrence Kaplan, in a recent study of the bean collection from this site, has reported that *Canavalia* (*C.* sp.) was present with *Phaseolus lunatus* from the early pre-ceramic levels through those of the ceramic-bearing Cupisnique Period. Macbride in his *Flora of Peru* (1943, vol. 13, pt. 3, no. 1, pp. 317–322) describes eight species of *Canavalia* but does not include either *C. ensiformis* or *C. obtusifolia.* His descriptions of the seeds of several of these species closely resemble the seeds found in pre-Columbian sites. The current revision of the genus by Jonathan D. Sauer will, in all probability, throw light on the identity of the *Canavalia* remains from Coastal Peru.

Erythrina *L.*

Erythrina is a genus of about thirty species of the tropics of both hemispheres. It comprises trees or shrubs with conspicuous flowers borne at a time when the leaves are lacking. The linear pods contain from one to many seeds. These are ellipsoid, dark, often marked with red, and with an elliptic to oblong hilum. Certain species of *Erythrina* are commonly planted for shade in cocoa and coffee plantations while others are used as ornamental trees.

Erythrina falcata *Benth.* in Mart. Fl. Bras. 15, pt. 1 (1859) 172.
Pisonay (Herrera, 1939, p. 282).

Trees of this species are reported to have been used as ornamentals in the plazas of towns in the Departments of Cuzco and Apurimac (Yacovleff and Herrera, 1935, p. 38).

Erythrina sp.

Yacovleff and Herrera (1935, p. 43) suggest that certain seeds found at Recuay and Pachacamac and described by Harms (1922, pp. 171–72) as possibly representing a species of *Ormosia* are more probably seeds of *Erythrina coralloides* Ses. et Moc. This latter species, however, is native to the West Indies and is not found in Peru. Other specimens of *Erythrina* in the form of whole and broken pods as well as loose seeds were found in Inca and Inca-associated levels at Pachacamac (Towle, 1948) (pl. II, 3), and similar specimens were among the remains recovered from the Decadent Maranga levels at Vista Alegre (Towle, 1958). These remains may represent *E. falcata* or a related species.

INDIGOFERA *L.*

Indigofera, a genus of about 275 species of the tropics of the Old and New Worlds, is the source of the indigo or añil of commerce, obtained primarily from two species, *I. tinctoria* of Asia and *I. suffruticosa,* a native of the American tropics.

Indigofera suffruticosa *Mill.* Gard. Dict. ed. 8 (1768) no. 2.
Indigofera anil L. Mant. (1771) 272.
Añil (Herrera, 1939, p. 113).—*Añil-añil,* Cuzco (*ibid.*)—*Mutui* (*ibid.*, p. 252).—*Mutui-cube* (*ibid.*).

Fester, in his studies of the pre-Columbian dyes of Peru, found that this species was used both alone and in combination with other colorants in the dyeing of cotton and wool and, in one instance, feathers (Fester and Cruellas, 1934; Fester, 1953). Yacovleff and Herrera, in their discussion of *Fourcroya andina* (1934, p. 268), mention the discovery of a mummy bundle at a site in the Chillón Valley that had false hair made of maguey fibers that had been dyed blue with añil. These same authors later (1935, p. 87) state that there is a tradition in the province of Convención, in the Department of Cuzco, that *Indigofera suffruticosa* was grown at an early time for the extract that produced the blue dye known as indigo or añil. Wittmack (1888, p. 347) states that the pre-Columbian Peruvians used indigo, obtained from a wild species ordinarily designated as *I. anil*, as a dye.

INGA *Willd.*

Inga is a large genus of about two hundred species of shrubs and trees of tropical and subtropical America. It is much used as a shade tree in coffee and cocoa plantations. The fruit consists of narrow, straight or somewhat curved pods, often flattened or four-sided, the margins frequently enlarged. The seeds of certain species are surrounded with a white, sweetish pulp that is used as food by man. The edible fruits of this genus are commonly called by the vernacular names *pacae* and *guaba* or variants of them.

Inga Endlicheri (*Kuntze*) *Macbride* Fl. Peruv. 13, p. 3, no. 1 (1943) 23.
Inga fasciculata Poepp. and Endl. Nov. Gen. et Sp. 3 (1845) 79.
Pacay amarillo (Macbride, 1943, vol. 13, pt. 3, no. 1, p. 23).

The pods of this Peruvian species contain a sweet edible pulp. Rochebrune (1879, pp. 346, 350) mentions *Inga fasciculata* and *I. Feuillei* Willd. as the two species of this genus that were used for food in pre-Columbian times on the coast of Peru. He further states that remains of the plant were fairly common at Ancón. It appears, however, that he had in mind not *Inga fasciculata* Willd. but *Inga fasciculata* Poepp. and Endl., which is now designated as *I. Endlicheri.*

Inga Feuillei *DC.* Prodr. (1825) 433.
Pacae (Herrera, 1939, p. 261).—*Pacay* (*ibid.*, p. 262).—*Paccai*, Cuzco (*ibid.*).—*Pa'qay*, Quechua (Mejía Xesspe, 1931, p. 11).—*Pa'qaya*, Aymara (*ibid.*).

Pacae is often grown for its shade on the coast of Peru and in the temperate valleys of the Sierras. It produces pods two to several dm. in length and about two cm. in breadth. These are flat with slightly enlarged margins, and the valves may show light striations on the external surfaces when dried. The seeds are surrounded by a white, sweet pulp that is much esteemed as a confection.

Numerous specimens from archaeological sites on the coast have been attributed to this species. Safford (1917a, p. 17) identified wooden poles used in the construction of graves at Ancón as *Inga Feuillei.* Two pods from this site are reported in the literature. One of these is illustrated by Wittmack (1880–1887, pl. 107, fig. 31); the other is described by the same author as having been found on the body of a child (1888, p. 328). Harms (1922, p. 169) mentions still another pod from Ancón and a *pacae* pod and numerous embryos contained in a gourd shell recovered from Chuquitanta (Harms, 1922, p. 169; pl. 1, fig. 4), which he identifies as presumably *Inga Feuillei.* Seeds and pods of this species were also found in Inca and Inca-associated levels at Pachacamac (Towle, 1948) (pl. II, 5), while seeds and remnants of pods occurred frequently in the excavations at Cahuachi in the Nazca Valley. Similar specimens, although few in number, were recovered from the neighboring sites of Estaquería and Huaca del Loro and from the burials at Ocucaje in the Ica region (Towle, 1956). Mangelsdorf (1942) lists *pacae* seeds among the plant remains that he identified from Paracas Necropolis. Leaves of *Inga* were used in the construction of mummy bundles at Ancón. According to Wittmack (1880–1887) leaves of two species of this genus, one of them *I. Feuillei,* were found. Harms (1922, p. 170), who studied the remains of *Inga* leaves from this site, identifies one as *Inga Feuillei;* the other remains undetermined. Yacovleff and Herrera (1934, p. 267) report that bodies surrounded by *pacae* leaves were found in sites in the Chillón Valley. Excellent reproductions in clay of *pacae* pods have been recovered from Santa Ana (Wiener, 1880, p. 601), Moche, Chimu, Ica (Yacovleff and Herrera, 1934, p. 267) and Chimbote (Safford, 1917a, p. 17, fig. 2) (pl. IX, B). Unidentified specimens of *Inga* were found with early pottery in Cupisnique levels at Huaca Prieta (Whitaker and Bird, 1949, p. 6).

LUPINUS (*Tourn.*) *L.*

Lupinus is a genus of several hundred species chiefly confined to the New World, with the greatest diversity occurring in California and Peru. The plants are herbaceous, semi-shrubby or shrubby, and may be either annual or perennial. Macbride lists eighty-eight species for Peru; *Lupinus mutabilis* is the only one cultivated. Weberbauer also gives this species as the only Peruvian cultigen.

Lupinus mutabilis *Sweet* Brit. Flow. Gard. 1, pt. 2 (1825) 130.
Lupinus "cunninghamii" of Cook, in Journ. Hered. 1925, p. 43.
Lupinus Cruckshanksii Hooker Bot. Mag. t. 3056.
Lupinus Tauris of authors, not of Benth. Pl. Hartw. 128.
Altramuz, Spanish (Herrera, 1941a, p. 14).—*Chocho,* Spanish name used in northern Peru (Weberbauer, 1945, p. 619).—*Tarhui,* name used in southern Peru (*ibid.*)—*Tarwi,* Quechua (Mejía Xesspe, 1931, p. 11).—*Tauri,* Aymara (*ibid.*).

Lupinus mutabilis is a tall, handsome annual with six to eight lanceolate leaflets arranged digitately and racemes of blue or white, sometimes yellowish-white, flowers. The fleshy pods are about eight cm. long and sixteen mm. broad and contain three to five seeds. These are either black, white or black and white. This species grows wild in the Andes from Colombia to Bolivia at altitudes from 800 to 11,000 meters. The edible seeds contain a bitter ingredient which necessitates their being either placed in running water for several days or soaked before they are cooked. In the latter instance, the first water that is poured off is used as an insecticide and as a fish poison. When grown in the lower altitudes this lupine is often a secondary crop in the same fields with maize, while in the higher altitudes it replaces the latter as a food plant. In 1915, Cook collected a lupine at Ollantaytambo, Department of Cuzco, Peru, known as "tarhui" by the natives who cultivated it for its edible seeds (U.S.D.A., 1917, p. 44). The plant was later (U.S.D.A., 1918) identified as *L. Cruckshanksii* Hooker. Subsequently, Cook (1925) mistakenly referred to this lupine as "probably *L. cunninghamii*," a name which does not appear in the literature.

Several writers during the Colonial Period in Peru mention the cultivation and use of this lupine which was known to the natives as "tarhui." However, there are no references to the finding of actual specimens in archaeological sites. Yacovleff and Herrera (1934, p. 395) do mention a Tiahuanco design, presumably from the southern coast, which they think is possibly a representation of the *tarhui* and which they designate as *L. Tauris*. Their description of the plant, however, is that of *L. mutabilis* and not *L. Tauris*. The latter is a small, bushy herb found in the mountains of Colombia and Ecuador. In fact, Herrera in a later publication correctly identifies the "tarhui" as *L. mutabilis*.

Lupinus sp.

Wittmack (1888, p. 338) reports that he found a few lupine seeds at Ancón; these were small and brown with darker brown spots. He dismisses the possibility of their being the "taroi" of which Garcilaso speaks, for the latter says that the lupine seeds that he observed in Peru were large and white. Wittmack leaves the seeds unidentified, since there are so many species occurring in Peru. Harms (1922, p. 172) quotes this reference from Wittmack and concludes that it is not possible to identify the seeds because of the small number of specimens available.

Mucuna *Adans.*

Plants of this genus, comprising about fifty species, are more or less woody vines found in the tropical areas of Asia, Africa and the Americas. The pods, which contain few seeds, are thick, linear to broadly ovate and are usually covered with bristles. The seeds of certain species of *Mucuna* are used medicinally.

Mucuna elliptica (*Ruíz et Pav.*) *DC.* Prodr. 2 (1825) 405.
Mucuna inflexa (Ruíz et Pav.) DC. Prodr. 2 (1825) 405.
Llamapañaui, "eye of llama" (Macbride, 1943, vol. 13, pt. 3, no. 1, p. 315).

This climbing plant grows wild in the forests of the tropical valleys of the Andes. The seeds are orbiculate, convex, slightly compressed and nearly surrounded by the hilum that is five to six mm. wide. Rochebrune (1879, pp. 346, 352) notes that the pulverized seeds of this species were used as a purgative by the ancient Peruvians. Among the plant remains that accompanied the mummy of a child that was found at Ancón was a large, pierced, spherical seed with a large hilum; Wittmack (1888, p. 327) thinks that this was part of a necklace, and he identifies the seed as *Mucuna*, probably *Mucuna inflexa* DC.

Mucuna sp.

An oval, pierced seed, practically surrounded by its wide hilum (six to eleven mm.) was found in the Inca level at Pachacamac (Towle, 1948). The cotton string upon which it had been strung was still attached (pl. IV, 1).

Myroxylon *L. f.*

Myroxylon is a small genus of trees represented by three species found in tropical America. The pods are compressed and have one seed at the tip; the sterile part is two-winged. A balsam-like substance in both the trunk and pods characterizes the genus. This material is used in medicine and as a fixative in the perfume industry.

Myroxylon sp.

Safford (1917a, p. 22) reports having found, in 1887, a lump of resin-like material included in a work basket that had been placed with a mummy at Ancón. Two additional specimens were recovered from the same site. One of these was a small spheroid gourd that had been split in two and filled with a black substance; the other was an oblong container (he identifies this as the fruit of *Crescentia*), the bottom of which held the resinous material (*ibid.*, p. 23, fig. 9). In each instance, this had become hard and brittle, resembling obsidian, and gave off a fragrant odor when rubbed. This substance he identifies as "balsam of Peru." There are two Peruvian species from which it could have been derived. One is *Myroxylon Balsamum*, the *Balsam of Tolú* or *estoraque* (Herrera, 1939, p. 181); the other is *M. peruiferum*. The former is found in Venezuela, Colombia and Peru; the latter extends from Argentina and Bolivia to Colombia and British Guiana. The balsam of Peru which Safford referred to is *M. Pereirae*, found from Brazil to Venezuela and in Central America. The common name is a misnomer, since it does not grow in Peru. Yacovleff and Herrera (1935, p. 67) refer the specimens that Safford describes to *Myroxylon peruiferum*. They do not give additional references to specimens of balsam.

Ormosia *Jacks.*

The approximately twenty-five species of *Ormosia* are found in the tropics of both the Old and New Worlds. The hard seeds, red, occasionally yellow and sometimes red and black, are often mistaken for those of *Erythrina*, another tropical genus.

Ormosia sp.

Wittmack (1888, p. 328) reports finding a seed of a leguminous plant in the collection of Peruvian archaeological materials made by Lieutenant Holland (site not given) which he identifies as *Ormosia*, "evidently O. *coccinia* Jacks.," a species reported from the Amazon basin in Peru and occurring in Brazil and the Guianas. Seeds of this genus are used for necklaces by the natives of the area (Levi-Strauss, 1950, p. 476). After quoting the above identification by Wittmack, Harms (1922, p. 171) describes similar seeds found at Recuay and Pachacamac. A group of seeds, some loose, some tied together with a string to form a necklace, were recovered from the former site. They are similar in shape and color to seeds found among the species of *Ormosia*. Three black seeds recovered from Pachacamac, he believes, might also belong to this genus. Yacovleff and Herrera (1935, p. 43), however, suggest that the seeds described by Harms might possibly be those of the genus *Erythrina*.

Pachyrrhizus *Rich.*

Pachyrrhizus, a genus which comprises about six species, is indigenous to Mexico, Central America and South America, where certain species were cultivated for their edible, tuberous roots before the coming of the Spanish. The beans are poisonous. The cotyledons contain a toxic resin which may prove to be of value as an insecticide (Clausen, 1945). Species of this genus are commonly referred to in England as "yam bean" and in Spanish as *jicama* or a variant. *Jicama* is believed to be derived from the Aztec word "xicama."

Pachyrrhizus Ahipa (*Wedd.*) Parodi An. Acad. Nat. Agron. and Vet. Buenos Aires 1 (1935) 137.
Ajipa (Parodi, 1926, p. 181).—Ahipa.

Ahipa is cultivated in northern Argentina and Bolivia, where it is said to have been grown since early times. The tuberous roots are oblong or spindle-shaped and light brown in color. The seeds are reniform, from eight to eleven mm. long, and dull black. Some botanists consider this species to be merely a cultivated variant of *Pachyrrhizus tuberosus*. A well-preserved tuber recovered from Paracas Necropolis has been assigned to this species (Mangelsdorf, 1942).

Pachyrrhizus erosus (*L.*) *Urban* Symb. Antill. 4 (1905) 311.
Pachyrrhizus angulatus Rich. ex DC. Prodr. 2 (1825) 402.
Chicam, Maya name evidently cognate with the Mexican *xicama*, which is derived from the Nahuatl (Standley, 1930, p. 299).

The natural distribution of this species appears to be northern Central America and Mexico, where it is still cultivated for its edible tubers. It is also widely cultivated and naturalized in the tropical regions of both the New and Old Worlds, and is grown at the present time on the north coast of Peru (Weberbauer, 1945, p. 619). This *jicama* is an herbaceous vine with tuberous roots, either turnip- or spindle-shaped. The fruits are from 7.5 to fourteen cm. long and contain either square or rounded seeds, five to eleven mm. long and five to ten mm. wide. They are usually slightly flattened and either yellow, brown or red in color. A small gourd, recovered from Chuquitanta, contained a necklace of round to egg-shaped, dark-colored beans. Harms (1922, p. 174), after careful study, concludes that these probably belonged to *Pachyrrhizus angulatus* (*P. erosus*), since the seeds of *P. tuberosus* which he examined were markedly different from those recovered from Chuquitanta. However, Yacovleff, after a thorough investigation (1933), assigns *P. tuberosus* rather than *P. erosus* to pre-Columbian Peru.

Pachyrrhizus tuberosus (*Lam.*) *Spreng.*, Syst. (1827) 281.
Asipa, Quechua (Harms, 1922, p. 176).—Villú, Aymara (*ibid.*).—*Xiquima* (*ibid.*).

Pachyrrhizus tuberosus, a climbing herb, is thought to be native to western South America, where it is found on the headwaters of the Amazon River and its tributaries in Brazil, Ecuador, Peru and Bolivia. The plant grows from a very large, subglobose, tuberous root. The rounded, reniform seeds are usually red, but occasionally black or black and white. They are sometimes shiny. Yacovleff has identified roots of this species from graves at Paracas (1933; Yacovleff and Muelle, 1934, p. 135; Yacovleff and Herrera, 1934, p. 283). Another root was also found among the plant remains in a mummy bundle from Paracas Necropolis (Towle, 1952b, pp. 235–36) (pl. I, 3). Whole plants of *jicama* as well as individual parts are often pictured in the art of Nazca and Paracas, and these have been assigned to *P. tuberosus* (Yacovleff, 1933; Yacovleff and Herrera, 1934, pp. 282–83; O'Neale and Whitaker, 1947, p. 320).

Phaseolus *L.*

A genus comprising approximately 170 species of the warmer regions of both hemispheres.

Phaseolus lunatus *L.* Sp. Pl. 724 (1753) 694.
Phaseolus Pallar Mol. Sagg. Chil. (1782) 130.
Pallar (Yacovleff and Herrera, 1934, p. 284).

The lima bean, originally a twining perennial, is a plant of varying habit. The pods are typically four to seven cm. long, with the upper margin recurved and the lower broadly rounded. The seeds are of varying color and both the pods and seeds are diverse in size and shape. There are climbing and also bush varieties which are treated in temperate climates as annuals and in tropical areas as perennials. Wild forms of the lima bean have been found in Guatemala, and a study of these, together with the cultivated forms, points to this area as the primary center of domestication of the species. Because of its great diversity authorities differ as to its classification. Mackie (1943), after an intensive study, considers the various forms as constituting one species, *Phaseolus lunatus*. He recognizes three distinct forms: the Hopi of Mexico and southwestern United States, the Carib of Central America, the West Indies and northern South America, and the Inca of Peru. Other botanists consider that two species are involved, *P. lunatus*, the small lima or sieva bean, and *P. limensis*, the large lima.

In the present study, the lima bean will be discussed according to Mackie's classification, namely, as the Inca form of *P. lunatus*. Both beans and pods of this type are characterized by their large size and are often referred to in the literature as *Phaseolus Pallar*. Archaeological evidence indicates the importance of the lima bean in the pre-Columbian culture of Peru. Wittmack (1880–1887, pl. 7, figs. 6–9) describes and illustrates beans from Ancón as *Phaseolus Pallar*. Costantin and Bois (1910, pp. 248–50, fig. 2) in their discussion emphasize that *P. Pallar* and *P. lunatus* are identical and remark that Wittmark inclined toward this conclusion at the time of their writing. Harms (1922, pp. 176–78) describes specimens from Chuquitanta, Paracas and Ica. Still other specimens have been recovered from Paracas Cavernas (Mangelsdorf, 1942), from the Inca and Inca-associated levels of Pachacamac (Towle, 1948), from a burial at the site of Zapollango (Peabody Museum, Harvard University), and from the Chillón Valley (Yacovleff and Herrera, 1934, p. 284). These latter authors, as well as Harms, emphasize the great variability that exists in the size and color of the beans they examined. From the south coast, well-preserved beans and pods of the bean occurred to a depth of 4.00 m. in the general excavation at Cahuachi (pl. II, 4) and at the site of Huaca del Loro in the Nazca Valley and in a burial at Ocucaje (Towle, 1956).

In his recent study of the bean collection from Huaca Prieta in the Chicama Valley, Lawrence Kaplan has found that lima beans occur among the specimens from the lower levels of Incipient Agriculture in the Huaca through the Cupisnique levels in the refuse heap. The lima is in association throughout with remains of *Canavalia* sp. and in the Cupisnique Period, also with the common bean, *Phaseolus vulgaris*. With one exception, the lima bean pods belong to the large-seeded or Inca type. The exception consists of four pods found in the upper levels where pottery, maize, peanuts and the common bean occur. These pods have characteristics similar to the *sieva* or small-seeded lima of Central America. It is of interest that limas showing these variations should occur in association with the ceramics and the additional cultigens of the Early Guañape and Cupisnique Periods.

The lima bean not only played an important part in the food economy of the people of the coast of Peru but in their art and symbolism as well. Design elements of beans and pods in the art of Nazca and Paracas have been attributed to *Phaseolus lunatus*. Some of the beans depicted are shown as germinated seeds with well-developed roots and, in some instances, leaves (Yacovleff and Herrera, 1934, p. 287, fig. 16; O'Neale and Whitaker, 1947, p. 320). One interesting design development from Moche shows processions of "lima-bean warriors" running in single file across the pattern (Yacovleff and Herrera, 1934, p. 288, fig. 17). Interspersed among these are realistic representations of limas. Such depictions have sometimes been interpreted as demonstrating that the pre-Columbian cultures had a well-developed ideographic system of reniform glyphs, which the lima bean with its varied marking and colors may possibly have inspired. These ideograms appear on ceramics and textiles from Nazca, Paracas, Tiahuanaco and Lambayeque (Larco Hoyle, 1946, p. 175).

Phaseolus vulgaris *L.* Sp. Pl. (1753) 723.

Frijol (word in various forms), Spanish name originally applied to beans of a related species.–*Haba,* common name in some localities of Peru. The Old World name for the broad bean, *Vicia Faba.*–*Mukulli,* Aymara (Mejía Xesspe, 1931, p. 11).–*Poroto* (Macbride, 1943, vol. 13, pt. 3, no. 1, p. 306).–*Purutu,* Quechua and Aymara (Mejía Xesspe, 1931, p. 11).

The common bean, also known as the bush, kidney, navy, string or snap bean, is characteristically a climbing annual with oblong pods that usually contain four to six seeds. Bukasov (1930) and his associates favor the highlands of Mexico and Guatemala as the primary center of domestication of the species. Added support is given this theory by the conclusions reached by Kaplan in his recent study of the cultivated beans grown in the southwestern United States during prehistoric times (1956). In Peru, this cultigen is usually planted in the temperate valleys of the coast and the highlands, and from the archaeological evidence, it appears to have been a staple food crop in pre-Columbian times (pl. II, 2).

Pods and separate valves of the common bean have been recovered from Cupisnique levels at the site of Huaca Prieta in the Chicama Valley. These specimens averaged 8.5 cm. in max. length and 1.5 cm. in max. breadth. The average measurements of the six seeds that were recovered were 1.1 cm x .5 cm. x .4 cm. (Towle, MS). According to Rochebrune (1879, p. 350), the largest number of seeds of *Phaseolus* recovered from the graves of Ancón were those of *P. stipularis* (a species the identity of which is in doubt); the remainder were *P. lunatus* and *P. multiflorus* (*P. coccineus*). Wittmack (1888, p. 334), however, believes that the specimens that Rochebrune attributed to both *P. stipularis* and *P. multiflorus* were nothing but the common bean, *P. vulgaris.* Rochebrune possibly had been influenced by the then popular belief that *P. vulgaris* was a native of Asia.

Pods of this species averaging twelve cm. in length were recovered from a site in the Chillón Valley (Yacovleff and Herrera, 1934, pp. 286–87). These contained from two to four elliptical seeds and were identified as *P. vulgaris ellipticus* Mns. Seeds of a similar shape were found at Paracas Necropolis, as well as slender ones that were assigned to *P. vulgaris ellipticus* and *P. vulgaris oblongus* Savi. These seeds vary in color from brown through purple to black, with a white hilum, and are marked with brown, yellow-brown or yellow-red-brown spots. They are often shiny. Specimens of *P. vulgaris* have been recovered from the sites of Cahuachi and Estaquería in the Nazca Valley (Towle, 1956). Reproductions of beans, probably representing *P. vulgaris,* appear on the pottery of Nazca (Yacovleff and Herrera, 1934, p. 289), and certain design elements from an embroidery of the Early Nazca Period have been identified as the common bean (O'Neale and Whitaker, 1947, p. 320). Approximately ninety seeds of this species were recovered from a grave at Huaca de la Cruz, a Mochica site, in the Virú Valley (Towle, 1952c, p. 355). Both pods and beans have been identified among the plant remains found in Inca and Inca-associated levels at Pachacamac (Towle, 1948). Wittmack (1880–1887, pl. 7) recognizes two varieties of the common bean at Ancón: *P. vulgaris oblongus* Savi., the long kidney bean or "date bean" (figs. 10, 11), and *P. vulgaris ellipticus* Mns., the elliptical kidney bean or egg-shaped bean (figs. 12, 13). Costantin and Bois (1910, pp. 245–46, fig. 1) illustrate and describe seeds of these two types. Safford (1917a, p. 13) recognizes two varieties of the common bean found in a jar of black ware, and again, four types included in a net bag (p. 16). However, he does not give any varietal names. Harms (1922, pp. 178–79) describes and identifies specimens from Ancón, Chuquitanta, Pachacamac and Ica. Many of these were in containers with other plant remains, such as lima beans, maize, seeds of *Cucurbita moschata,* peanuts, fruits of *Campomanesia* and *Erythroxylon* as well as a pod of *Caesalpinia corymbosa.* Because of the great diversity in the color and size of the common beans found in archaeological materials, Harms does not think it advisable to attempt to identify them further as to variety, an opinion in which the writer concurs.

Phaseolus sp.

Chui, Tchui (and other variants), Quechua (Yacovleff and Herrera, 1934, 289).

Yacovleff and Herrera (*ibid.*) describe a wild form of *Phaseolus* that grows in the tropical river valleys of the mountains. Its round, varicolored beans are not edible, but are used in the playing of games. They also describe a seed of this plant that was found among a number of beans of *P. vulgaris* from Paracas. This round bean, nine mm. long, nine mm. broad and seven mm. thick, was red in color and marked with black. A similar seed was recovered from a site in the vicinity of Lambayeque. Yacovleff and Herrera believe that these specimens represent the types of beans that are mentioned by Cobo, Garcilaso and Morua as having been used by the Peruvians during the Colonial Period in playing games. Safford (1917a, p. 16) reports finding smooth, round beans at Ancón that he believes to be distinct from *P. vulgaris.* He identifies these as the "tchui" beans of the Quechua Indians and as the ones referred to by Cobo. Inca games in which beans were employed are described by Rowe (1946, pp. 288–89).

PROSOPIS *L.*

This genus comprises about thirty tropical and subtropical species of the Old and New Worlds. These plants, either shrubs or trees, produce linear pods with compressed, albuminous seeds and often with a sweetish, edible pulp. The best known species, the mesquite, *Prosopis juliflora,* occurs in the southwestern desert region of the United States, Mexico, Central America, the West Indies and northern South America. A variant of this species is often credited to Peru in the literature, but it does not occur in that country. The Peruvian equivalent is *P. chilensis,* a plant very similar to *P. juliflora* and often confused with it.

Prosopis chilensis (*Mol.*) Stuntz U.S. Bur. Pl. Ind. Inventory 31 (1914) 85.
Prosopis juliflora of authors as to Peru, not (Sw.) DC.
Thacco, Quechua (Macbride, 1943, vol. 13, pt. 3, no. 1, p. 109).–*Algarroba* (*ibid.*).–*Huarango,* Ica (Herrera, 1939, p. 197).

The *algarroba,* which grows abundantly on the coast and in the valleys of the western sierras, yields a gum that is used similarly to gum arabic. Its wood is a source of charcoal. The sweetish, edible pods (five to fifteen cm. long) and the ground beans constitute a valuable livestock food. There are historical references in the literature that mention the seeds of *algarroba* as having been used for food by man in pre-Columbian times. Designs on pottery from Moche have been identified as *P. juliflora* DC. The wood of this tree has been found in the tombs of Nazca and other sites on the southern coast (Yacovleff and Herrera, 1934, pp. 291–92), and was used with canes in the constructions at the site of Huaca Prieta de Guañape during the Cerro Prieto Period (Strong and Evans, 1952). *Algarroba* pods and seeds have been recovered from Cahuachi and Huaca del Loro in the Nazca Valley (Towle, 1956). Pottery jars have been found that appear to have been mended with a resinous material resembling the black gum secreted by this tree (Yacovleff and Herrera, 1934, p. 292). In Wittmack's discussion of "Beans" (1888, p. 338) he mentions finding fragments of a string of beads and strongly veined flat pods that he identifies as *Prosopis glandulosa,* now considered to be a variant of *P. juliflora.* Similarly, although *algarroba* pods found by Safford (1917a, p. 17) in graves at Ancón were identified by him as merely *Prosopis* sp., a more correct determination would be *P. chilensis.*

OXALIDACEAE. Oxalis Family.

Oxalis *L.*

A genus of more than 250 species of the tropical and temperate regions of both hemispheres.

Oxalis tuberosa *Mol.* Sagg. Chil. 3 (1782) 109.
Oca (or *occa*) (Hodge, 1946, p. 214).–*O'qa,* Quechua (Mejía Xesspe, 1931, p. 11).–*O'qa, apina,* Aymara (*ibid.*).

The *oca,* a small, erect, branching, herbaceous plant, bears trifoliate leaves and yellow-to-orange-colored flowers, the petals of which are sometimes marked with purple lines. The slender, edible roots have a smooth skin bearing numerous scales which cover long, deep eyes. This plant is cultivated from Bolivia to Colombia as a minor root crop, considered of less importance than the potato but more important than *ullucu* and *añu*. It is grown on terraces in the *quebradas* of the cold sierras and in the altiplano zone to an altitude of 3,800 to 4,000 meters. There are several forms of *oca*: the bitter, called *cjaya-oca* (Hodge, 1946, p. 216) which has white tubers, and the sweet, the tubers of which are of various colors. The native names (*ibid.*) which designate the varieties are *sapallu-oca* (yellow to orange, with red margined scales), *chachapea-oca* (grayish), *pauccar-oca* and *lluchcho-oca* (reddish), and *mestiza-oca* (white).

Oca tubers contain calcium oxalate crystals, which are especially abundant in the bitter variety. This requires the tubers to be cured in the sun before they are eaten. The sweet form is sometimes made into a product known as "cavi," in the preparation of which sun-dried tubers are cooked slowly and then eaten with cane syrup or honey as a dessert. *Chuño* is prepared from the bitter variety of *oca* in essentially the same way that it is made from the potato. This product furnishes a staple winter food for the Indians in the Andean region in which *oca* is grown.

Yacovleff and Herrera (1934, pp. 307–309) state that remains of tubers of *oca* have not been recovered from sites on the coast. However, one specimen of a small dried tuber was among the remains recovered from Inca and Inca-associated levels at Pachacamac (Towle, 1948) (pl. I, 1). A Tiahuanaco design from Pacheco which closely reproduces a plant of this species is illustrated by Yacovleff and Herrera (1934, fig. 28d). They also include a Chimu vessel of black pottery (*ibid.*, fig. 23e) made in the form of a tuber with numerous molded "eyes." This they likewise identify as *Oxalis tuberosa.*

TROPAEOLACEAE. Nasturtium Family.

TROPAEOLUM *L.*

Tropaeolum, the only genus in the family *Tropaeolaceae*, is represented by about forty-five species occurring from Mexico to Chile, chiefly in the cooler regions. It is best known for the species and hybrids which are the familiar nasturtiums of gardens throughout the world.

Tropaeolum tuberosum *Ruíz et Pav.* Fl. Peruv. 3 (1802) 77 pl. 314.
Añu, Quechua, most widely used in southern Peru (Hodge, 1946, p. 220).—*Apina-mama* (*ibid.*).—*Isaña*, Aymara, used in the Titicaca basin (*ibid.*).—*Mashuar*, used from Central Peru to Ecuador and southern Colombia (*ibid.*).

Plants of this species have glabrous, twining stems with round, five-lobed leaves. The flowers are orange to red in color and are smaller than those of the garden *Tropaeolum*. The edible cone-shaped tubers resemble those of *oca* (*Oxalis tuberosa*) in both form and color. The color variants are recognized by a number of native descriptive names, among them *yana-añu* (black), *puca-añu* (red), *yurac-añu* (white), *sapallu-añu* (yellow) and *Muru-añu* (spotted) (Macbride, 1949, vol. 13, pt. 3, no. 2, p. 619). *Añu* is grown as a cultigen in the high Andes but it is the least popular of the four major root crops of the region. The tubers are not palatable when raw and must be "cured" by the sun or made into *chuño*. This product resembles the *chuño* made from potatoes, *ullucu* or *oca*.

The Indians in early times are said to have attributed anti-aphrodisiac properties to the tubers when eaten, while the inhabitants of the Andes today believe they possess medicinal attributes. Yacovleff and Herrera (1935, p. 47) reproduce a Tiahuanaco design of an entire plant of *Tropaeolum tuberosum* (Yacovleff and Herrera, 1934, fig. 28c) that was recovered from Pacheco.

ERYTHROXYLACEAE. Coca Family.

Erythroxylon *P. Br*

The genus *Erythroxylon,* native to tropical and subtropical South America, consists of about ninety species. Several of these are widely cultivated for their leaves which contain the economically important alkaloid, cocaine. The two most important species are *E. Coca* and the closely related *E. novogranatense.* The former is found in the warm regions of Peru, Argentina, Bolivia and Brazil, usually at high altitudes, while the latter, although having the same general distribution, occurs at lower elevations.

For centuries, the use of coca leaves as a stimulating masticatory has been widespread among the peoples of the Amazon basin. By the time of the rise of the Inca Empire its use had become well established in the Central Andes. Coca is one of the recognized economic plants of that period, and certain plantations were set aside for its cultivation. It is of interest that some of these are still used for this purpose. The strange properties possessed by the leaves were attributed to divine origin, and thus its use was originally restricted to the Incas and members of the nobility. However, the coming of the Spaniards and the ensuing disintegration of Inca society placed the use of coca leaves at the disposal of the common people. With them its popularity has continued, until today it is almost universally used.

The modern method of chewing coca is similar to that used in pre-Columbian times. The dried leaves are carried in a small bag; a wad of leaves is chewed and formed into a small quid, which is held against the cheek. To this is added lime or the ash of certain plants, on the assumption that alkaline qualities are necessary to free the alkaloid in the leaves. Although the necessity for this is questioned by some modern authorities, anyone who has ever chewed coca knows he gets no effects without the addition of an alkali. The lime or ash (*llipta*) is either formed into pellets or used as a powder. The latter is carried in a small container (*poporo*) and dipped out with a stick or small spoon.

The wide use of coca in pre-Columbian Peru is reflected in the archaeological specimens recovered from many coastal sites. Both Nazca and Mochica jars show men with distended cheeks or dipping lime from *poporos* (Yacovleff and Herrera, 1934, p. 297) and the finding of both bags of leaves and lime pellets have been recorded. The latter are described by Harshberger (1898, p. 4) as being three inches long and one and one-half inches thick, hardened and cracked by drying. The fact that the coca plant entered into the art of these early times is manifested by Wiener's statement (1880, pp. 49, 55) that leaves of this genus were reproduced in gold and silver, and again by Yacovleff and Herrera's comment (1934, p. 297) that a collar made of stone beads carved to resemble the seeds of this plant has been found. Specimens of coca leaves that have been identified as to species are mentioned below.

Erythroxylon Coca *Lam. Encycl.* 2 (1786) 393.
Coca (Herrera, 1939, p. 150).–*Cuca,* in Cuzco (*ibid.,* p. 158).

The coca, a leafy shrub with reddish-brown bark, ordinarily grows to a height of one to two meters. Its many, slightly erect branchlets have elliptic-obovate opposite leaves measuring four to seven cm. in length and three to four cm. in width. They are acute at the base, and although they may be slightly so at the tip, are obviously pointed. The mid-nerve, which is impressed above but slightly conspicuous beneath, is transversed on the upper surface by a light ridge that is very prominent on the underside. In addition, many leaves have two conspicuous curved lines that run longitudinally on either side of the mid-nerve about one-third of the distance between it and the leaf margin. The small white flowers produced in clusters in the axils of the leaves and branchlets develop reddish-orange drupes, two to eight mm. long and 3.5 to four mm. thick. These are oblong-ovoid in shape, pointed at the tip and three-angled with deep, longitudinal furrows.

Bags filled with well-preserved leaves of coca have been recovered from Ancón (Saffray, 1876, p. 402; Wittmack, 1888, p. 328). Other bags containing the remains of leaves of *E. Coca,* powdered calcium or ashes of quinoa, and leaves of *Ilex paraguariensis* are described by Rochebrune (1879, pp. 346, 352–53). He also cites leaves of *E. rigidulum* among these particular plant remains. However, this species is native to the mountainous areas of Colombia and Venezuela, and since the identification was made on the strength of the leaves alone, it should be accepted with extreme reservation. Coca, presumably *E. Coca,* has been reported from a tomb at Paracas Cavernas (Bennett and Bird, 1949, p. 142), and a poorly preserved quid of leaves included in a mummy bundle from Paracas Necropolis, almost certainly belongs to this species (Towle, 1952b, p. 243). An unusually fine textile pouch filled with coca leaves and drupes was found in a tomb of the Coastal Tiahuanaco Period at Vista Alegre (Towle, 1958) (pl. IV, 5). Leaves of *Erythroxylon Coca* were incorporated in the symbolism of Inca art (Herrera, 1923, p. 445).

Erythroxylon novogranatense (*Morris*) *Hieronymus* in Engler's Bot. Jahrb. 20, Beibl. no. 49 (1895) 35.

This species is closely related to *E. Coca* and may prove to be only a variety of it. It differs mainly in having smaller leaves which are blunt at the tip. Harms (1922, p. 180) mentions two bags of coca leaves from Pachacamac, one from Ica, and two small bags from Chuquitanta, which he identified as those of *E. novogranatense*. Since the leaves of this species are so similar to those of *E. Coca*, Harms suggests the possibility that other specimens of coca leaves that have been identified as being *E. Coca* may be those of *E. novogranatense*. Harms further identifies small fruits found at Ancón by Reiss and Steübel as this species.

ZYGOPHYLLACEAE

Porlieria *Ruíz et Pav.*

Porlieria is a small genus of three species found in Mexico and Andean South America. The hard, tough wood of the Peruvian species is said to resemble that of the related "lignum vitae" (*Guaicum* sp.) and is used in the manufacture of small objects such as spoons and combs.

Porlieria hygrometra *Ruíz et Pav.* Syst. Veg. (1798) 94.
Murucho (Macbride, 1949, vol. 13, pt. 3, no. 2, p. 653). This is the word for "hard" in Quechua (Markham, 1864, p. 147).

The wood of certain articles found in coastal sites has been identified as of this species. According to Rochebrune (1879, pp. 346, 354), sticks and weaving implements were carved of *Porlieria hygrometra*. Saffray (1876, p. 401) identifies the head of a club found in a mummy bundle as "Gaiacwood." Since *Guaiacum* is not found in Peru, he apparently is speaking of the similar wood of *Porlieria*.

MALPHIGIACEAE. Malphigia Family.

Bunchosia *Rich.*

The forty or more species of this genus are mostly native to Peru and Brazil, with a few in Mexico and the Antilles. Nine species are described for Peru (Macbride, 1950, vol. 13, pt. 3, no. 3, pp. 855–61), but because of the variability within the genus they are often difficult to distinguish.

Bunchosia armeniaca (*Cav.*) *Rich. apud Juss.* in Ann. Mus. Par. 18 (1811) 481.
Ciruelo del fraile (Weberbauer, 1945, p. 620).—*Ciruela del fraile.*

Bunchosia armeniaca is a tree four to twelve meters in height, with leaves twenty-four cm. long and eleven cm. broad. The flowers, which are borne on axillary racemes, produce edible fruits in the form of two-seeded drupes, usually

2.5 cm. long and two cm. in diameter. It is cultivated in Peru at the present time. Tello (1938, p. 108) illustrates three Mochica jars decorated with small molded fruits which he identifies as "Ciruelas del fraile, *Bunchosia armeniaca.*" (pl. XII). Seeds of a fruit, identified as those of *ciruelo del fraile,* were found in the pre-ceramic levels of Huaca Prieta and were attributed to *Bunchosia* sp. (Whitaker and Bird, 1949, p. 3). Similar specimens have been recovered from Inca and Inca-associated levels at Pachacamac (Towle, 1948) (pl. III, 1) and at two other central coast sites, namely Vista Alegre and Playa Grande (Towle, 1958). In all probability, these are also seeds of *Bunchosia armeniaca.*

EUPHORBIACEAE. Spurge Family.

Manihot *Adans.*

Manihot is a South American genus of about one hundred species, several of which are a source of rubber. Others are widely cultivated for their edible, starchy, tuberous roots. There is much variability and research has not yet determined the taxonomic significance of the variable characters.

Manihot esculenta *Crantz* Inst. Herb.: 1 (1766) 167.
Jatropha Manihot L. Sp. Pl. (1753) 1007.
Manihot utilissima Pohl Pl. Bras. Icon. 1 (1827) 32, pl. 24.
Cassava.–*Manioc.*–*Rumu,* Aymara and Quechua (Mejía Xesspe, 1931, p. 11).–*Yuca,* a name derived from the island Arawack.–*Yuca amarilla* (Macbride, 1951, vol. 13, pt. 3A, no. 1, p. 171).–*Yuca blanca* (*ibid.*).

The slender stems of this variable, shrubby perennial, usually from two to three meters high, rise from clustered roots ending in elongated tubers. The palmate leaves are divided into three to seven lobate sections and the marble-spotted seeds are contained in a capsule with six longitudinal wings. This plant is said to be native to Brazil but it is now widely cultivated. In Peru it grows at altitudes between 1,500 and 2,200 meters, and it is also reported on the coast from Arica northward (Sauer, Carl O., 1950, vol. 6, p. 508). The varieties of this cultigen are usually designated as bitter (poisonous) or sweet (non-poisonous) manioc. The tubers of the former contain sufficient prussic acid to require leaching or heating in order to make them edible.

Tubers of *Manihot esculenta* have been found in many sites on the coast, and they are recognized as having been a staple food in prehistoric Peru. Evidence suggests that manioc was known during the Cupisnique Period on the north coast (Larco, 1946, p. 150; Bennett and Bird, 1949, p. 126). Wittmack (1880–1887, pl. 107, fig. 14) records finding the root of *yuca* at Ancón, and Costantin and Bois (1910, p. 265, fig. 15) describe a narrow, spindle-shaped tuber which they identify as belonging to this species. Other specimens of *yuca* have been recovered from Chuquitanta, Pachacamac, Ica (Harms, 1922, p. 180), and Chillón and Paracas (Yacovleff and Herrera, 1934, p. 273). In addition, a tuber was identified among the remains in a mummy bundle from Paracas Necropolis (Towle, 1952b, p. 236). Manihot roots were also found in levels at Cahuachi (Plate I, 4), Huaca del Loro and Estaquería in the Nazca Valley (Towle, 1956.) Besides the actual specimens, there are many representations of both clusters and single tubers in clay and on textiles. These are reported from Nazca and Chimbote (Safford, 1917a, fig. 12, p. 14, 24; Yacovleff and Herrera, 1934, p. 272; O'Neale and Whitaker, 1947, p. 319). The Nazca pottery designs mentioned and identified by Yacovleff and Herrera (1934, p. 270) as those of *yuca* might equally well be representations of *arracacha.*

AQUIFOLIACEAE. Holly Family.

ILEX *L.*

Ilex is a large genus of about three hundred species found in the tropical and temperate regions of the Old and New Worlds. These plants are evergreen or deciduous trees and shrubs with alternate leaves and round or elliptical drupes. Many species are often cultivated for their handsome foliage and fruit. The leaves of several South American species, chiefly *Ilex paraguariensis* St. Hilaire, are the sources of the tea-like beverage *maté*, *yerba maté* or Paraguay tea. This plant grows wild in southern Brazil, Paraguay and Argentina. Safford (1917a, pp. 26–27) states that *Ilex paraguariensis* has not been reported from Peruvian archaeological sites although small, characteristic gourds called *maté*, from which the beverage is consumed and from which it gets its name, have been recovered from pre-Columbian graves in Argentina. Rochebrune (1879, pp. 345, 353) mentions finding leaves of *Ilex paraguariensis* with those of coca. This determination, however, is open to doubt. More probably, the leaves are those of a Peruvian species, several of which may be used satisfactorily for preparing *maté.*

SAPINDACEAE. Soapberry Family.

SAPINDUS *L.*

The genus *Sapindus* consists of about twelve species native to the tropics of Asia and America.

Sapindus Saponaria *L.* Sp. Pl. (1753) 367.
Soapberry–*Cholloco* (Herrera, 1939, p. 175).–*Choloque* (*ibid.*).–*Sullucu*, Cuzco (*ibid.*, p. 327).

This American species is a small tree varying from four to ten meters in height. The leaves consist of about four or five pairs of oblong, elliptical or lanceolate leaflets. The small, white flowers produce round, brown berries with orange-brown translucent flesh. Each contains one smooth, globular, black seed about 1.2 cm. in diameter. These are sometimes used for beads and buttons. The fruits, as well as other parts of the plant in lesser degree, contain the glucoside, *saponin*, which yields a soapy lather in water and forms an emulsion with fats and oils. It is this property that makes the fruits valuable commercially as a soap substitute. Both berries and seeds were used by the Peruvians in prehistoric times, the former presumably for cleansing, the latter for beads. Rochebrune (1879, pp. 346, 351) mentions finding whole fruits in bags at Ancón, although he considers them a source of food. A fruit of this species found at Ancón is shown by Wittmack (1880–1887, pl. 7, fig. 20). In his description of the uses of the fruit of *Sapindus Saponaria*, he says that the pulp was used as a soap and the seeds were worked into necklaces (1888, p. 348). Both fruits and seeds were found in Inca and Inca-associated levels at Pachacamac (Towle, 1948), while fruits and leaves were recovered from a site in the Chillón Valley (Yacovleff and Herrera, 1935, p. 40). From the south coast, several specimens of the fruit and seeds of *S. Saponaria* were recovered from the sites of Cahuachi and Huaca del Loro in the Nazca Valley (Towle, 1956) (pl. IV, 6).

ELAEOCARPACEAE. Elaeocarpus Family.

Aristotelia *L'Herit.*

A genus of about seven species native to the warm areas of the southern hemisphere.

Aristotelia Macqui *L'Herit.* Stirp. Nov. (1784), 31, t. 16.
Maqui (Record and Hess, 1943, p. 145).

This shrub, a native of Chile, grows to a height of about two meters, producing shiny, evergreen leaves up to eight cm. in length. The berries are eaten both fresh and preserved and are also used to color wine. Rochebrune (1879, pp. 346, 351) mentions the fruit of this species as having been occasionally used for food in prehistoric times at the site of Ancón.

MALVACEAE. Mallow Family.

Gossypium *L.*

The genus *Gossypium* consists of fifteen species and several varieties, both wild and cultivated, which are distributed in the tropical and warm temperate regions of the world. These annual and perennial shrubs, which are sometimes tree-like, have palmately ribbed or lobed leaves and large white, yellow or purplish flowers. The fruits are capsules which contain small, subglobose or angular seeds, which in all species of *Gossypium* are covered with some type of hairs. These vary greatly in size and abundance; in fact, in some species, they are so inconspicuous as to be easily overlooked. In the wild cottons, the seed hairs are short, firmly attached to the seed, and, in most species, circular in cross-section and unconvoluted; they are not spinnable. The hairs of the cultivated cottons, on the other hand, are of two types: long, thin, fine, flat, convoluted, loosely attached, spinnable hairs commonly known as lint, and short, coarse, firmly attached "fuzz" which forms an under-coat on the seed surface.

Only four species of *Gossypium* are cultivated; two of these (*Gossypium arboreum* and *G. herbaceum*) are natives of the Old World and two (*G. hirsutum* and *G. barbadense*) originated in the corresponding tropical and temperate areas of the New World. The origin of the New World cottons has been a subject of major interest to the geneticists since it was discovered that these two New World species of *Gossypium* are allopolyploids. That is to say, they are hybrids which were differentiated by a doubling of the chromosome number of a cultivated cotton of Asia and a wild cotton of the New World. The parent species of the American cottons are considered to be the wild *G. Raimondii* (or a similar ancestor) of northern Peru and one of the cultivated Asiatic cottons, probably *G. arboreum.* How this union of two such widely separated species was produced has caused and is still causing much speculation and controversy among botanists, geographers and archaeologists. The two present-day New World cultivated species are thought to have been derived in ancient times from this common ancestor, which developed in the northern Andean region. Of the two species, *G. barbadense* is of particular concern in the present paper. Its prehistoric range of distribution was almost wholly western South American; its present area of dispersal meets that of *G. hirsutum* in lower Central America and the West Indies. *G. peruvianum,* an Andean variant, is no longer considered to be a distinct species.

Gossypium barbadense *L.* Sp. Pl. (1753) 693.
Gossypium peruvianum Cav.
Algodon (Spanish).—Cotton—*Utcju* (Herrera, 1939, p. 348).—*Utcu* (*ibid.*).

Cotton was the most commonly used vegetable fiber in prehistoric Peru; the early chroniclers state that it was used extensively by both the Indians of the coast and those living in the hills. Archaeological sites from all sections of the coast have yielded specimens of *Gossypium,* including bolls, seeds, raw fiber, yarn, thread and a variety of manufactured materials. There are many references to cotton in the archaeological literature and also a large bibliography on the use of this fiber in the textile industry of the prehistoric peoples of the area. Among the characteristic cultural elements of pre-Columbian Peru are the elaborate and abundant woven articles found in the burial sites. These were made of cotton alone or in combination with fibers of *Furcraea* and the wool of llama, alpaca and vicuña. These prehistoric cotton fibers have been described as either white, tawny or vicuña-colored, or reddish brown. Colored cottons are cultivated in Peru at the present time.

The earliest evidence of cotton recovered from an archaeological Peruvian site is reported by Bird (1948b, p. 24) from Huaca Prieta, a pre-ceramic site located in the Chicama Valley on the north coast. The cotton fibers which were found were twisted into string without the use of spindle whorls and used in the making of fish nets, pouches and twined fabrics. After examining this material, Hutchinson (1959) has concluded that the Huaca Prieta cotton represents one of the earliest cultivated *barbadense* cottons. A gourd bowl filled with cotton fiber was recovered from a grave at the Aspero site at Supe (Towle, 1954, p. 135), and Rochebrune (1879, pp. 345, 354) speaks of having identified woven cotton fiber as well as raw cotton from Ancón. A number of cotton seeds were recovered from Grave 13, at Huaca de la Cruz in the Virú Valley (Towle, 1952c, p. 355). Cotton bolls, raw fiber and seeds were found in the Early, Playa Grande and Inca-associated levels at Pachacamac (Towle, 1948), and a gourd filled with cotton from Chuquitanta is described by Harms (1922, p. 181). Cotton fibers were used to wrap small objects and Costantin and Bois (1910, p. 257) found maize cakes wrapped in this fashion. In addition, cotton was used in various ways in the preparation of the mummy bundles found at Paracas (Towle, 1952b). Cotton seeds and bolls were recovered from Cahuachi and Huaca del Loro in the Nazca Valley (Towle, 1956).

PAVONIA *Cav.*

Pavonia is a genus of herbs, shrubs and small trees, native to the tropical and warm temperate areas of the Old and New Worlds. Its light, soft, odorless wood is similar to that of the related species, *Hibiscus.*

Pavonia paniculata *Cav.* Diss. 3 (1787) 135, t. 46, f. 2.

According to Rochebrune (1879, pp. 347, 354), this South American tree furnished the soft wood used for making decorated wooden cups, handled spoons, wooden ear ornaments and carved wooden cones. These latter were possibly used for ornamentation. He further says that objects made of *Pavonia paniculata* were quite common at Ancón.

BOMBACACEAE. Silk-Cotton Tree Family.

The family *Bombacaceae* comprises some twenty genera and 150 species, native to the tropics of both hemispheres. Many of these woody plants are characterized by large capsules or pods, the small seeds of which are surrounded by a mass of fine hairs. In some species, these fibers are the source of a filling material. Much confusion has arisen as to the identity of the several genera involved. Both Saffray (1876, p. 402) and Rochebrune (1879, pp. 345, 355) identify the fibers used for stuffing dolls and for wrapping small objects as the floss of *Bombax Ceiba.* This species was earlier attributed to the tropical regions of Asia, but recently has been described as native to America. However, the plant is not reported from Peru. Harms (1922, p. 181), considering *Bombax Ceiba* as a native of Asia, suggests that the fibers examined by Saffray and

Rochebrune were those of *Ceiba pentandra*, a species of a related genus and the source of the kapok of commerce. Again, this species, although a native of America, like *Bombax Ceiba* is not found in Peru. Two other bombacaceous species, however, *Bombax Ruizii* and *Chorisia insignis*, are found in northern Peru and the valleys of the central Andes. Both of these produce floss, that of the latter closely resembling the fibers of *Ceiba pentandra*. It is quite probable that the fibers examined by Saffray and Rochebrune were from one or the other of these two plants.

STERCULIACEAE. Cocoa-Tree Family.

THEOBROMA *L.*

The genus *Theobroma* consists of about twenty species, including both wild and cultivated forms. These are native to the marginal tropical regions of Mexico, Central America and northern South America. The plants are rather small trees from five to thirteen meters in height with alternate, simple leaves. The small flowers are borne on short stems in either auxiliary clusters or directly on the trunk and large branches. The fruits are woody, podlike capsules fifteen to twenty cm. long, tapering at both ends and filled with a mucilaginous pulp that surrounds the numerous seeds. It is from the seeds of certain species that chocolate (from the Nahuatl *chocolatl*) is obtained. *Theobroma Cacao* L. is the most widely cultivated form. It was grown on the Pacific slopes of Mexico and Central America in pre-Spanish times, and we have early descriptions of its preparation as a food and as a beverage in that area. The wild ancestors of this important cultigen have not been definitely determined. It appears that chocolate was unknown in Peru before the Conquest, since there are no references to its use in the early accounts of the useful plants of Peru nor have there been actual remains of the plant recovered from archaeological sites. Consequently Wiener's statement (1880, p. 601) that the fruits of cacao and possibly the seeds were reproduced in prehistoric pottery is open to question.

CARYOCARACEAE. Souari Family.

CARYOCAR *L.*

A small genus of ten species found in tropical America. The wood of these plants is used for making furniture and the oily seeds are roasted and eaten.

Caryocar amygdaliferum *Cav.* Ic. 4 (1797) 37 tt. 361, 362.
Almendras de chachapoyas (Yacovleff and Herrera, 1935, p. 57).—*Almendro* (Herrera, 1939, p. 108).—*Tampa* (Yacovleff and Herrera, 1935, p. 57).

The almond-like seeds of the drupe of this species have an agreeable taste and are esteemed as food. Although a native of Colombia, the plant was reported by Cobo (1890) as native to the mountainous area of Chachapoya in northern Peru. Wittmack (1888, p. 341) says that the fruit was rarely found in pre-Columbian graves, but Safford reports that actual specimens were recovered from Ancón (1917a, p. 21; fig. 6). He further describes and illustrates a pottery vase from Chimbote molded in the form of two enlarged bean-shaped kernels of the seed (fig. 7).

OCHNACEAE. Ochna Family.

Godoya *Ruíz et Pav.*

Godoya is a genus consisting of three species native to the American tropics.

Godoya obovata *Ruíz et Pav.* Syst. Veg. (1798) 101.

This shrub or small tree grows in the forest of the eastern slopes of the Andes at altitudes between 1,300 and 1,800 meters. Rochebrune (1879, pp. 346, 354) states that specimens of lance handles and small clubs made of the hard, compact wood have been commonly found in sites on the Peruvian coast.

BIXACEAE. *Annatto Family.*

Bixa *L.*

A monotypic genus native to the American tropics.

Bixa Orellana *L.* Sp. Pl. (1753) 512.

Achihuiti (Herrera, 1939, p. 102).–*Achiote* (*ibid.*).–*Achote* (Weberbauer, 1945, p. 621).–*Annatto.* Name applied to the dye extracted from the pulp about the seed, as well as to the tree.–*Mantur* Quechua (obsolete) (Yacovleff and Herrera, 1934, p. 300).

Bixa Orellana comprises bushy shrubs or small trees with entire heart-shaped leaves. The panicles of white or pink flowers produce bivalve capsules covered with spines. The small, ovoid, pointed seeds are surrounded by a thick, red-brown pulp. This readily dissolves in water, leaving upon evaporation a cake of orange pigment, the source of the commercial dye known as "annatto" that is used in coloring foods, such as cheeses, butter and butter substitutes. The natives living in the area where *Bixa Orellana* is indigenous have utilized this vegetable pigment as a body paint. In Peru, this species grows in tropical and subtropical areas up to 1,800 and 2,000 meters altitude, but apparently it is not found wild. Wittmack (1880-1887) describes a square pottery vessel found in a burial at Ancón which contained both the dried pulp and the seeds of this plant. This container was made with four divisions, each covered with a tied cloth. Three of these compartments were filled with the dried seed pulp while the fourth held a number of whole seeds. He further gives *Bixa Orellana* as the source of a yellow dye in pre-Columbian Peru (1888, p. 328, 347). Specimens of the spiny pods of *annatto* were among the vegetal remains from Cahuachi on the south coast (Towle, 1956).

FLACOURTIACEAE. Flacourtia Family.

PINEDA *Ruíz et Pav.*

A genus of two species found in the tropical and subtropical areas of South America.

Pineda incana *Ruíz et Pav.* Prodr. (1794) 76, pl. 14.

This low shrub, often with a number of stems, has slightly oblong leaves which are covered with a white pubescence. The leaves are said to supply a black dye (Macbride, 1941, vol. 13, pt. 4, no. 1, p. 29) and the strong wood of the plant is utilized for making stools, baskets and in the construction of hanging bridges as well. This species is found in Ecuador and Peru where it grows on the eastern and northwestern slopes of the Andes and in the inter-Andean valleys. Rochebrune (1879, pp. 345, 354) states that both large and small clubs and lance handles made of the hard wood of *Pineda incana* have been recovered from burial sites on the coast.

PASSIFLORACEAE. Passion-Flower Family.

PASSIFLORA *L.*

The species, approximately 350 in number, which comprise this genus are chiefly native to the tropical and subtropical regions of the New World. They are herbaceous or woody tendril-bearing vines, usually with solitary, showy purple, white or greenish-yellow flowers. The fruit, commonly known as passion-fruit or *granadilla*, is an indehiscent berry up to three decimeters in length. The interior contains a mucilaginous pulp which surrounds the numerous seeds. Seventy species are described for Peru (Macbride, 1941, vol. 13, pt. 4, no. 1, p. 91), the genus being widely dispersed in areas up to four hundred meters in altitude. Several species are known for their edible fruits, among which are *P. ligularis* Juss., known by the Aymara name of *apincoya* (Cook, 1925, p. 99; Mejía Xesspe, 1931, p. 11) and *P. mollissima* (HBK) Bailey, the common name of which is *tumbo* (Weberbauer, 1945, p. 620). This latter name is also applied to fruits of other species (Herrera, 1939). The *granadillas* were used for food in pre-Columbian Peru. Yacovleff and Herrera (1935, pp. 32-33) describe three species and Wittmack (1888, p. 342) refers to the plant as the source of an edible fruit. However, these authors do not refer to any archaeological remains or reproductions of the plant that have been identified.

CARICACEAE. Papaya Family.

CARICA *L.*

Carica is a small genus of more or less succulent trees native to the American tropics. The palmately-lobed leaves form a crown at the summit of the straight trunk, while the yellow, melon-like fruits are borne on long stems just below this leafy cluster. One species, *C. Papaya,* is widely grown for its excellent fruit. Although this plant is sometimes cultivated in Peru, it is unknown in the wild state and there is no positive proof that it was grown in prehistoric times. However, some pottery reproductions suggest the papaya, and Wiener (1880, p. 601) mentions this fruit as one that was used as a model for clay vessels. A number of other species of *Carica* are native to Peru, among them *C. candicans* (*mito, jerju, uliucana* [Weberbauer, 1945, p. 180]), a plant found on the western slopes of the Andes and in the *lomas* region. The fruit of this tree, and that of several other species to a lesser degree, is used widely as food, and it might be that the pottery reproductions are of one of these fruits rather than that of *C. Papaya.*

CACTACEAE. Cactus Family.

Several genera of the Cactus Family characteristic of the xerophytic areas were important in the ethnobotany of the pre-Columbian cultures of Peru. Some species furnished edible fruits or spines utilized as needles and pins, and cacti were often depicted in designs and plastic forms. In northern Chile, long cactus thorns were used extensively for fishhooks (Bird, Correspondence, 1957). It is, however, often difficult or impossible to identify the species. The more common genera were *Cactus, Cereus, Lobivia, Opuntia* and *Trichocereus.* The Chroniclers recognized two types of cacti, *tunas* and *pita-hayas,* the division being made on whether or not the corolla was persistent in the fruit (Yacovleff and Herrera, 1934, p. 317).

CACTUS *L.*

Cactus sp.

Globular stems and fruits reproduced in certain Mochica pottery designs have been identified by Yacovleff and Herrera (1934, p. 320) as those of a cactus of this genus, designated by them, however, as *Melocactus.*

CEREUS *Miller*

Cereus sp.

Many of these plants are characterized by tall, upright stems which are generally branched. The elongated, funnel-formed flowers are nocturnal and the fleshy fruits of some species are edible. A number of design units painted on Mochica pottery clearly represent the vegetative parts of some of the cacti of this genus (Yacovleff and Herrera, 1934, fig. 38).

LOBIVIA *Britton and Rose.*

Lobivia is a genus of about twenty species, all of which are found in the highlands of Peru, Bolivia and Argentina.

Lobivia corbula *Britton and Rose* Cactaceae, 3 (1922), 56.
Anapancu, Cuzco (Herrera, 1939, p. 113).—*Hachcana*, Aymara (Yacovleff and Herrera, 1934, p. 319).

This species grows on the hills of the high Andean valleys of Peru. The plants are globular in form with twelve or more ribs separated by deep, longitudinal indentations. The areoles are covered with white wool and the spines, in clusters of six to nine, are yellow. The flowers are salmon-red and produce a sweet, edible fruit. This cactus is cultivated as an ornamental plant. Yacovleff and Herrera (1934, p. 319, fig. 37) give a reproduction of a Nazca vase molded in the form of a globular cactus which they identify as possibly this species, to which, however, they give the name *Mammillaria Herrerae.*

OPUNTIA *Tour. ex Mill.*

The genus to which the prickly pear belongs numbers about 250 species. It is one of the largest and most widespread genera in the family, ranging in North America from Massachusetts to British Columbia and southward to the Straits of Magellan.

Opuntia exaltata *Berger* Hort. Mortol. (1912) 410.
Ppata-quisca, Cuzco (Herrera, 1939, p. 283).—*Espino*, Cuzco (*ibid.*, p. 180).

Plants of this species are tree-like with a definite trunk which grows from two to five meters high. They are many-branched with rounded areoles filled with short, white wool. The spines are numerous, sometimes twelve or more in a cluster and often thirteen cm. long. They are brown in color and have a roughened tip. The flowers are brick-red and the fruits are green and pear-shaped. This cactus is used for fences about small farms and is also cultivated for its spines, which are used as pins. It occurs in Ecuador, Peru, Bolivia, and probably northern Chile. In Peru, it is well established in the western hills and along the coast, and is probably a native of Cuzco, where it is cultivated extensively. Combs and pins made of the spines of *Opuntia exaltata* have been frequently found in coastal sites (Yacovleff and Herrera, 1934, pp. 317–18). An insect, *cochinilla*, lives on this plant, and since cochineal, the red dye produced from this insect, has been identified from archaeological sites on the coast, it is probable that this cactus or a closely related species was employed in raising the insects in this area (*ibid.*).

Opuntia sp.

Cactus stems comprised of oval and oblong joints were painted on pottery from Moche. These can be identified as stems of a species of *Opuntia* closely resembling *O. ficusindica*, a cactus that was not in Peru in pre-Columbian times (Yacovleff and Herrera, 1934, figs. 38, f, h).

Opuntia spp.

Species of *Opuntia* are widely distributed in Peru. Harms (1922, p. 182) says that plants of this genus are often reproduced on pottery vessels.

TRICHOCEREUS *Riccob.*

A genus of about seventeen species found in South America.

Trichocereus cuzcoensis *Britton and Rose* Cactaceae, 2 (1922). 136.
Giganton, Cuzco (Herrera, 1939, p. 186).—*Jahuackollai*, Cuzco (*ibid.*, p. 210).

The stems of this tall, columnar cactus are five to six meters high, much branched and somewhat spreading. The stout, rigid spines are numerous, often twelve in a cluster. The plant produces a substance known as *nopal* gum, which is used in Cuzco at the present time as a mucilaginous additive to whitewash. The fruits are edible. This cactus grows not only in the highlands but also in the *quebradas* and on the sandy beaches of the coast. Both the vegetative parts of the plant and the fruits frequently appear as elements in painted designs on Mochica pottery. This is particularly true in hunting and battle scenes (Yacovleff and Herrera, 1934, p. 320).

LYTHRACEAE.

LAFOENSIA *Vand.*

This genus of about ten species is found in the American tropics.

Lafoensia punicaefolia (*Bert.*) *DC.* in Mem. Soc. Phys, Host. Nat. Geneve 3, pt. 2 (1826), 86, pl. 1.

These trees, thirteen to sixteen meters high, have oblong, lanceolate leaves and bell-shaped flowers arranged in simple racemes. The smooth capsules are ovoid, ending in a sharp pointed tip. This plant and another South American species, *L. pacari*, produce a beautiful yellow dye obtained from the leaves and young twigs. Rochebrune (1879, pp. 346, 255) states that the leaves of *L. acuminata*, a Peruvian species, were the source of dye in pre-Columbian Peru. A mixture of these leaves with those of *Piper lineatum*, *Dicliptera Hookeriana* and *D. peruviana* (the source of blue and yellow dyes) gave a green color. A capsule of a tree of this genus was recovered at Ancón and was identified by Wittmack (1880–1887, pl. 106, figs. 18, 19) as *L. acuminata.* Later, Harms (1922, p. 182) described nine whole fruits and a number of seeds of *Lafoensia* which he found at Chuquitanta. He says, however, that neither his specimens nor that recovered from Ancón and described by Wittmack could be *L. acuminata,* since the upper part of the fruit of that species is radiately star-shaped, while those recovered by himself and Wittmack were smooth. These latter specimens appear to be identical with fruits of *L. punicaefolia,* a tree which Harms says is not reported for Peru. Macbride, however (1941, vol. 13, pt. 4, no. 1, p. 219) gives its distribution from Bolivia and Peru to Venezuela and Guatemala.

MYRTACEAE. Myrtle Family.

Campomanesia *Ruíz et Pav.*

Campomanesia is a South American genus of about eighty species of trees and shrubs.

Campomanesia lineatifolia *Ruíz et Pav.* Fl. Peruv. 4 (1802), t. 422.
Palillo (Herrera, 1939, p. 264).

Trees of this species are found in the Andean region from Chile to Colombia, and are cultivated for their edible fruit in the tropical regions of the Peruvian Andes. These small, round fruits closely resemble those of *Psidium Guajava.* They are crowned by a persistent calyx of five-parted, irregular lobes, and the thin skin surrounds a soft, granular pulp which encloses the small seeds. Although the flavor of the fruits of *C. lineatifolia* is often stronger than that of the guava, they are much used for making preserves and condiments.

Rochebrune (1879, p. 347) says that the fruits of this species, although used for food at the present time, are rarely found in archaeological sites. However, Harms (1922, p. 182) identifies several small ball-shaped specimens that were in a container with other plant remains recovered from Ancón as fruits of this species. In addition, he describes two small fruits found in a gourd at Chuquitanta, one of which he illustrates (*ibid.* pl. 1, fig. 11). Several specimens were recovered from Pachacamac (Towle, 1948), while well-preserved fruits of *C. lineatifolia* were frequently found at Cahuachi in the Nazca Valley (pl. III, 4). The species was only sparsely represented at the neighboring site of Huaca del Loro (Towle, 1956).

Psidium *L.*

Psidium is a tropical American genus of about one hundred species of trees and shrubs.

Psidium Guajava *L.* Sp. Pl. (1753) 470.
Psidium pyriferum L. Sp. Pl. ed. 2 (1763) 672.
Guava.–*Guayabo.–Sahuinta,* Quechua (Markham, 1864, p. 168).–*Sawintu,* Quechua and Aymara (Mejía Xesspe, 1931, p. 11).

Both wild and cultivated forms of the guava are found from Mexico to Peru. These slender trees, of a height up to ten meters, have opposite, simple leaves which are veined above and pubescent beneath. The yellow fruits vary in form from globular to pear-shaped and measure 2.5 to ten cm. in length. The white, yellow or pink flesh is soft and granular. The numerous, small seeds are three to five mm. long and kidney-shaped or triangular.

An oblong, compressed fruit of *P. Guajava* was recovered from Ancón (Wittmack, 1880–1887, pl. 106, fig. 17). Whole, globular fruits were found in Gallinazo and Mochica levels in the Virú Valley (Castillo de Tomaval; Huaca de la Cruz) (Towle, 1952c, p. 355), as well as in the Inca levels at Pachacamac (Towle, 1948). Fruits were also obtained at Cahuachi and Estaquería in the Nazca Valley and from the sub-surface level at the Ica site at Ocucaje (Towle, 1956) (pl. III, 3). Although Yacovleff and Herrera (1934, p. 274) state that no representations of this species have been observed in the art of pre-Columbian Peru, according to O'Neale and Whitaker (1947, p. 319) the fruits are depicted on pieces of Early Nazca embroidery.

MELASTOMACEAE. Melastoma Family.

ACISANTHERA *P. Br.*

This genus of low-growing plants consists of twenty species found in the American tropics.

Acisanthera inundata *Triana* in Trans. Linn. Soc. 28 (1871) 33.
Microlicia inundata DC. Prodr. 3 (1828) 117.

These plants are trailing, recumbent herbs which grow in moist locations in Brazil and Guiana. Rochebrune (1879, pp. 346, 354) reports that the fibrous, flexible stems were used for some of the cording, basketry and mats often found in simple burials on the Peruvian coast. Harms (1922, p. 183) questions this identification because of the absence of the species in Peru.

MICONIA *Ruíz et Pav.*

A large genus of about six hundred species of the American tropics.

Miconia chrysophylla (*L. C. Rich.*) *Urb.* Symb. Ant. 4 (1910) 459.
Melastoma fulva L. C. Rich. in Bonpl. Melast. (1816) 23, pl. 11.
Miconia fulva DC. Prodr. 3 (1828) 180.
Puca-mullaca (Herrera, 1939, p. 284).

This shrub is characterized by the ashy to reddish scale-like pubescence that covers the young branchlets and the underside of the leaves. The wood is used locally for interior construction (Record and Hess, 1943, p. 357). Rochebrune (1879, p. 346) gave this plant as the source of a dye and mentioned that evidence of its use by the ancient Peruvians had been occasionally encountered. Harms (1922, p. 182) doubted this statement since he thought this tree was not found in Peru. Both Macbride (1941, vol. 13, pt. 4, no. 1, p. 398) and Williams (1936, p. 386), however, record its presence in northeastern Peru.

ONAGRACEAE. Evening Primrose Family.

Fuchsia *L.*

The genus *Fuchsia* consists of about sixty species native to Central and South America, with a few species reported from New Zealand.

Fuchsia boliviana *Carr.* in Rev. Hortic. (1876) 150.

Coapac-Ñucchchu; in Quechua, the first word means "superior"; the latter is the name for the *Salvia*, another flower held sacred by the Inca (Yacovleff and Herrera, 1935, p. 86).—*Chimpu-chimpu,* Quechua *(ibid.)*.—*Uchu-uchu* (Herrera, 1941b, p. 323).

Fuchsia boliviana is a branching evergreen shrub two to three meters in height, with oval to elliptical leaves. The handsome, red-scarlet, elongated flowers are borne on short, axillary racemes, the long stamens protruding from the corolla. The fruit is an oval, purplish berry which has a sweetish taste and is said to possess mild narcotic properties (Yacovleff and Herrera, 1935, p. 85). This species is reported from Bolivia and the Department of Cuzco, where it grows from 1,800 to 3,000 meters altitude. Like the salvia and *cantua*, *Fuchsia boliviana* was considered a sacred flower by the Incas and probably was used by them in religious and political ceremonies. The flowers in a design on an Inca *kero* have been identified as probably those of this plant *(ibid.*, fig. 63).

UMBELLIFERAE. Parsley Family.

Arracacia *Bancr.*

A genus of stout, perennial herbs representing about thirty species native from Mexico to Peru.

Arracacia xanthorrhiza *Bancr.* in Jamaic. Jour. 4 (1826) 18.

Arracacia esculenta DC. Prodr. 4 (1830) 244.

Apio, Spanish word for celery and given by them to *A. xanthorrhiza* because of its celery-like flavor (Cook, 1925, p. 99); (for *Apium graveolens* (celery) Herrera, 1939, p. 114).—*Arracacha* (Herrera, 1939, p. 115).—*Ra'kacha*, Aymara (Mejía Xesspe, 1931, p. 11).—*R'qacha*, Quechua *(ibid.)*.—*Virracas*, Spanish (Herrera, 1942b, p. 184).

The *arracacha,* a perennial herb, two to three feet high, has deeply divided leaves and loose umbels of small flowers. It produces large, fleshy roots that have an agreeable flavor and constitute an important food item among the people of Central America and Andean South America. This plant is cultivated in the temperate regions of the highlands of Colombia, Ecuador, Peru and Bolivia. It is reported by the chroniclers to have been grown in the same general area at the time of the Conquest. Safford (1917a, p. 24) identifies roots, which are reproduced as part of a pre-Colombian vase from coastal Peru, as possibly those of either *Arracacia edulis* (a Mexican species) or *Xanthosoma.* He includes "arracacha roots" among the food plants in an exhibit which is illustrated in Plate II. Hodge (1954, fig. 17) points out that certain Nazca pottery design units representing clusters of roots that have been ascribed to *yuca* (Yacovleff and Herrera, 1934, p. 270) might as well be those of the root portion of *arracacha.*

LOGANIACEAE. Logania Family.

Buddleia *L.*

The seventy or more species comprising this genus are native to the tropical and subtropical regions of Asia, America and to a limited extent, South Africa.

Buddleia sp.

Quisuar, and its variants (Weberbauer, 1945, p. 621; Herrera, 1939, p. 296; Yacovleff and Herrera, 1935, pp. 41–42).

The plants of this genus are shrubs or occasionally herbs, with opposite, lanceolate leaves and four-lobed white, violet to purple flowers borne in heads, panicles or spikes. Weberbauer (1945) reports fifteen species of *Buddleia* for Peru, several of which are grown in the temperate regions as ornamentals and for their wood, which is used in tool-making and cabinet work. Herrera (1923, pp. 445–46) states that the wood of *Buddleia incana* had symbolic meaning and was used ceremonially by the Incas. Although Yacovleff and Herrera (1935, pp. 41–42) do not mention this species, they do describe two other species of *Buddleia* (*B. longifolia* and *B. coriacea*), both of which furnish wood for tools and construction. Under *B. longifolia,* these authors further call attention to a pre-Columbian temple at Cuzco, known as *Quishuar-Cancha,* in which was kept a gold idol the size of a ten-year-old boy, and say that idols burned at the festivals of *R'aimi* were also made of the wood of *Buddleia longifolia.* A number of small branches, several of which had the remains of dried, opposite leaves still attached, were found in a grave of the Classic Maranga Period at Vista Alegre. These have been identified as probably *B. americana,* a species mentioned by Weberbauer (1945) as growing on the Playa of the coastal region of Peru (Towle, 1958).

SAPOTACEAE. Sapodilla Family.

ACHRAS *L.*

A monotypic genus.

Achras Zapota *L.* Sp. Pl. (1753) 1190.
Sapota Achras Miller Gard. Dict. ed. 8 (1768) no. 1.
Sapodilla.–Chicle.

The *sapodilla,* a tall, evergreen tree, is indigenous to southeastern Mexico, but is widely cultivated in the American tropics for its edible fruit. The bark contains a milky juice which is the source of the commercially important chicle, the basis of the chewing gum industry. Wittmack (1888, pp. 342–43) reports finding at Ancón a necklace made of the glossy, dark brown shells of the seeds of the fruit of *Achras Zapota.*

LUCUMA *Mol.*

Lucuma is a large genus of shrubs and trees, several of which produce edible fruit. Most of the species are native in the American tropics. Some authors consider *Lucuma* and the closely related genus, *Pouteria,* to be synonymous, giving the older generic name, *Pouteria,* to the group. However, other writers, although referring many species of *Lucuma* to *Pouteria,* consider certain Andean species to be distinctive enough to be retained in a separate genus and so continue to recognize *Lucuma.*

Lucuma bifera *Mol.* Sagg. Chil. (1782) 187.
Lucuma obovata HBK. Nov. Gen. et Sp. 3 (1818) 241.
Lucuma (Herrera, 1939, p. 221).– *Lucumo* (Yacovleff and Herrera, 1934, p. 269).–*Lu'kuma,* Aymara (Mejía Xesspe, 1931, p. 11).–*Lu'qma,* Quechua (*ibid.*).–*Rujma* (Yacovleff and Herrera, 1934, p. 269).

Lucuma bifera, an evergreen tree, eight to ten meters high, has spreading branches with entire, elliptic-ovate leaves. The globose to ovate fruits, seven to ten cm. in diameter, are green in color and have a bright orange to yellow, dry, mealy pulp. The smooth skin is marked at the base by a persistent calyx and toward the apex by a ring of wrinkled tissue. The seeds, contained in separate loculi, are one to five (usually two) in number. They are subglobose in shape and are a glossy, light brown color, resembling a horse chestnut. The ball-like endosperm separates easily into two uneven, convex cotyledons. This species is a native of Peru and is cultivated for its edible fruit. The fine, compact wood is much esteemed for cabinet work.

Specimens of whole and halved fruits of this species, as well as seeds and cotyledons, are among the most common plant remains found in archaeological sites on the Peruvian coast (pl. III, 5). *Lucumas* were an important part of the diet of the ancient Peruvians, and halved fruits are frequently found in *petacas* with remains of other food plants. Harms (1922, pp. 183–84) describes a bag containing coca leaves, cotton and a piece of *lucuma* fruit from Ancón, and Wittmack (1880–1887, pl. 107) describes other fruits and seeds from this site. He originally identified these as probably *L. splendens* or *L. valparadisaea*, but he later assigned them to *L. obovata* (*L. bifera*). Several specimens of *lucuma* were recovered at Chuquitanta (Harms, 1922, p. 184). A cache of approximately one hundred halved fruits was found at a site in the Chillón Valley (Yacovleff and Herrera, 1934, p. 271) and parts of seeds (with *Canavalia* beans) have been reported from Paracas (*ibid.*). Remains of the fruit were also found by Bird in the pre-ceramic level at Huaca Prieta in the Chicama Valley (Whitaker and Bird, 1949, p. 3). Several whole and halved fruits and numerous seeds and seed parts were found at Pachacamac in Plain-Ware, Inca-associated and Inca levels (Towle, 1948). Seed fragments were recovered in a Gallinazo site and a whole fruit and seeds from a Mochica burial (Castillo de Tomaval, Huaca de la Cruz), both in the Virú Valley (Towle, 1952c, p. 355). *Lucuma* seeds and a few fruits were found throughout the excavated area at Cahuachi and at Huaca del Loro in the Nazca Valley (Towle, 1956).

The fruit is also a familiar motif in pre-Columbian art, represented in clay and in designs on both pottery and textiles (pl. XIII, B). Reproductions of *lucumas* in plastic form have been found at Moche, Lambayeque, Nazca and other sites on the coast, in addition to painted designs on pottery from Nazca and Paracas. These latter showed the characteristic calyx and wrinkled tissue at the apex and also the large hard seeds within the pulp (Yacovleff and Herrera, 1934, fig. 9, p. 271). The *lucuma* fruit is reported (O'Neale and Whitaker, 1947, p. 319) as an element of design on an Early Nazca embroidery. Wittmack (1888, pp. 328, 341) mentions that the leaves of this tree were sometimes found with mummy burials.

APOCYNACEAE. Dogbane Family.

THEVETIA *L.*

A genus of seven or eight species native to the American tropics.

Thevetia peruviana (*Pers.*) *Schum.* Eng. and Prantl Pflanzenf. pt. 4, 2 (1895) 158.

Thevetia nereifolia Juss. ex Steud. Nomencl. ed. 2, 2 (1840) 680.

Yellow oleander.—*Maichill,* in northern Peru (Yacovleff and Herrera, 1935, p. 88).—*Siatica,* in the vicinity of Lima (*ibid.*).

Plants of this species are evergreen shrubs or trees up to ten meters in height, with narrow, dark green, linear leaves from seven to fifteen cm. long. The fragrant flowers are a waxy yellow and funnel-like. These produce small, irregularly-shaped drupes, each containing a woody, triangular stone or endocarp which holds two seeds. All parts of this plant are poisonous, due to the presence of *thevetin,* a substance with digitalis-like qualities. The seeds are collected and used in the manufacture of medicine for heart conditions.

Safford (1917a, p. 23) states that the hollow stones of *Thevetia peruviana* were used by the people of pre-Columbian Peru for rattles and were sometimes attached to the fringe which was sewed on the border of ceremonial garments. He gives a picture of a stone recovered from a grave at Ancón. Necklaces made of similar specimens were also found at this site (Wittmack, 1880–1887) and these stones were used with some frequency as a decorative element on the Peruvian coast. Yacovleff and Herrera (1935, p. 88) reproduce a Mochica jar, the shoulder of which is decorated with a design containing these triangular stones in relief.

CONVOLVULACEAE. Morning Glory Family.

Ipomoea *L.*

The genus *Ipomoea,* with about four hundred species, some of which produce fleshy lateral roots, is native to both the tropical and the temperate regions of the Old and New Worlds. The best known species is the sweet potato, *I. Batatas.*

Ipomoea Batatas (*L.*) *Poir. ex Lam.* Encycl. 6 (1804), 14.
Convolvulus Batatas L. Sp. Pl. (1753) 154.
Batatas edulis Choisy in Mem. Soc. Phys. Geneve 6 (1845) 435.
Apichu, Quechua for the sweet variety (Herrera, 1942b, p. 183). Name also used by the Aymara. May have been borrowed from the Quechua or reverse may have occurred (Hawkes, 1947, p. 220).–*Batatas,* Caribbean word. –*Camote,* corruption of Nahuatl *camotl.*–*Cjumara* ('Kumara), Quechua for the starchy variety (Herrera, 1942b, p. 184).–*Tuctuca,* Aymara. May have been borrowed from the lowland tribes, since the Aymara live at altitudes above the limit of *Ipomoea Batatas* cultivation (Hawkes, 1947, p. 220).

The sweet potato is a long-trailing perennial vine with leaves five to fifteen cm. in length and of varying shape, even on the same plant. The colors of the funnel-form flowers are rose-violet, bluish or sometimes purplish with a darker center. This species, although a native of tropical America, is cultivated in the warmer regions of both the Old and New Worlds for its starchy, fleshy roots. It is propagated by the division of either the plant or the root, the latter having adventitious buds rather than eyes or scars. There are two varieties: one with dry, mealy, yellow flesh; the other, erroneously called yam, with a softer, sweeter, glutinous flesh which is deep yellow or orange in color. The sweet potato is grown in Peru on the coast and in the warm, inter-Andean valleys. Both varieties are cultivated and produce roots of various forms and colors. These are often designated by local, descriptive names.

Actual roots of *'kumara* have been found at Ancón (Wittmack, 1880–1887, Pl. 107, fig. 18; Safford, 1917a, p. 24) and similar specimens were identified from Inca and Inca-associated levels at Pachacamac (Towle, 1948). Tello (Nordenskiold, 1931, vol. 9, p. 51) reported having recovered a root of *Ipomoea Batatas* at Paracas, and it has been recently collected by Strong from Cahuachi in the Nazca Valley (Towle, 1956). Three other specimens of the sweet potato were among the vegetal remains from Ocucaje in the American Museum of Natural History collections (pl. I, 5). Yacovleff and Herrera (1934, p. 274) state that there are numerous specimens of Mochica pottery in the Museo Nacional at Lima which reproduce the roots of this plant, and they particularly mention one specimen from a site in the Chillón Valley.

POLEMONIACEAE. Polemonium Family.

CANTUA *Juss.*

The distribution of this small genus of about seven species is confined to the mountainous areas of Peru, Bolivia and northern Chile.

Cantua buxifolia *Lam.* Encyc. (1783) 603.
Ccantu, and its variants, in the Department of Cuzco (Herrera, 1941b, p. 347). —*Coelmo,* a name applied to the related species, *C. Pyrifolia* (*ibid.*).—*Flor del Inca* (Cobo, 1890, p. 483).

These plants are branching shrubs or small trees from two to three meters in height, with opposite, single leaves of a lustrous green. Since pre-Columbian times, *Cantua,* the present national flower of Peru, has been grown as an ornamental because of its brilliant, solitary, tubular flowers of red and yellow. These were considered to be sacred by the Incas, and they were often reproduced in their pottery designs (Herrera, 1923, p. 443; Yacovleff and Herrera, 1935, fig. 56, pp. 59–61). The Aymara name *Inca-panccara* (*Flor del Inca*) conveys the idea of the reverence in which the flower was held. Among the Indian population of Cuzco at the present time, the flowers are used in funeral rites; a branch placed before a house announces the selling of *chicha.*

LABIATAE. Mint Family.

SALVIA *L.*

The five hundred or more species that comprise this genus are native to the tropical and temperate regions of both hemispheres.

Salvia tubiflora *Smith* Ic. Ined. (1798) 26, t. 26.
Salvia biflora Ruíz et Pav. Fl. Peruv. 1 (1798) 24, t. 38.
Ñucchchu (Yacovleff and Herrera, 1935, p. 85).

This herbaceous annual grows profusely on the slopes of the coastal hills and in the mountains of Peru, particularly in the Departments of Lima, Arequipa and Cuzco. The plant, which attains a height of two to three dm., has the four-sided stems characteristic of the Mint Family, and opposite, oval-to-oblong leaves. The showy, axillary flowers have a persistent calyx and a fused corolla of blood-red petals. The species is widely cultivated as an ornamental plant. Cobo (1890, Lib. IV, cap. II, p. 338), in his account of the plants of the New World, mentions the salvia but does not further identify it. Although there are a number of species of the genus in Peru, Yacovleff and Herrera (1935, p. 85) describe only *Salvia tubiflora*, which undoubtedly had ethnobotanical significance at the time of Cobo as it does at present. Representations of the flowers often appear on Inca wooden cups (*keros*) and pottery. An infusion of the flowers and the flowers themselves were used by the Incas during their rituals conducted to placate the God of Earthquakes. This custom even now persists among the Indian population of Cuzco (*ibid.*; Herrera, 1923, p. 443).

SOLANACEAE. Nightshade Family.

Capsicum L.

Capsicum, a small genus, chiefly native to the American tropics, includes the various red peppers, known as *aji* (from the island Arawak) in Spanish South America (Sauer, Carl O., 1950, p. 521), *uchu* (Quechua) in Peru (Markham, 1864, p. 182) and *chile* (a Nahuatl derivative) in Mexico and Central America (Sauer, Carl O., 1950, p. 521). The plants are semi-woody or woody, annual or perennial shrubs with simple and entire leaves and small white or greenish-white flowers, sometimes tinged with violet to purple. Certain species are grown for their pungent fruits, which are widely used for food and condiments. The many-seeded, fleshy, pod-like berries vary greatly in size and shape among those species under cultivation. Because of the number of varieties of *Capsicum* that have been found in Guiana and Brazil, it has been suggested that this area is the possible home of the genus. However, further study is necessary before the place of origin can accurately be determined.

There is much confusion in the taxonomy of the genus. Opinions range from grouping all the cultivated peppers under one or two species with several varieties in each to giving specific rank to most of these varieties. Final assignment of types must depend upon a reappraisal of the phylogeny and genetic relationship of the members of this widely used cultigen. The following account includes four of the five species recognized by Smith and Hieser (1957; Heiser and Smith, 1953) in their recent studies of the cultivated peppers. All of the archaeological remains of *Capsicum* that have been identified as to species have been placed in one or another of these four.

Capsicum annuum *L.* Sp. Pl. (1753) 188.
El aji largo (Yacovleff and Herrera, 1934, p. 279).

Practically all of the larger fruited varieties of the present-day cultivated peppers grown in temperate areas belong to this species. Both the vegetative characters and the fruits vary greatly. The pods may be from one to thirty cm.

in length and range in shape from small cones to fruits that are squat or flattened with thickened flesh. In the immature stage, they are either yellow or green, turning to red-yellow and brown when ripe. It appears that the pepper was widely used during prehistoric times on the coast of Peru, although the archaeological specimens show little variation (pl. IV, 7). Rochebrune (1879, pp. 347, 351) recognized the fruits of this species among the plant remains from pre-Columbian Peru. He states that they were fairly common and although usually broken, had retained their color. Two varieties of *C. annuum* are described by Harms (1922, pp. 184–85; pl. 1, fig. 12) from pods contained in a gourd bowl recovered at Chuquitanta. His identification is based upon the shape of the fruit and characteristics of the calyx. One group of pods was narrow, five to nine cm. long, and pointed, with the calyx adhering closely to the base; these specimens he assigns to *C. annuum acuminatum* Fingerh. The second group comprised broader fruits, 3.5 to four cm. in length, 1.5 to 2 cm. in breadth, with a cylindrical shape tapering slightly toward the tip. The broad calyx barely touched the surface of the pod. Harms describes these as apparently *C. annuum abbreviatum* Fingerh. Specimens of peppers reproduced on pottery from Nazca, as well as actual fruits from the Chillón Valley, have been assigned to *C. annuum* (Yacovleff and Herrera, 1934, p. 279). This species has also been reported from among the plant remains found in two mummy bundles from the south coast, one from Paracas Cavernas (Mangelsdorf, 1942), the other from Paracas Necropolis (Anon, 1938, pp. 119-25).

Capsicum frutescens *L.* Sp. Pl. (1753) 189.

Capsicum frutescens is widely cultivated today in the tropical and subtropical regions and is used extensively in the diet of the people of Mexico, Central America and South America. The fruits of this plant vary greatly in both size and shape, but none seems to exceed ten cm. in length. Safford mentions five specimens of this species which were found in a pre-Columbian grave near Lima (1917a, p. 22). Pods depicted on an Early Nazca embroidery have been identified as those of *C. frutescens L.* because of their curved, pointed pods with an enlarged, persistent calyx (O'Neale and Whitaker, 1947, p. 320).

Capsicum pubescens *Ruíz et Pav.* Fl. Peruv. 2 (1794) 30.

Lo'koti, Aymara (Mejía Xesspe, 1931, p. 11).–*R'occoto* (Herrera, 1941b, p. 361).–Rocoto (Mejía Xesspe, 1931, p. 11).–Ro'qote, Quechua (*ibid.*).

This species is one of the two mentioned by Rochebrune (1879, pp. 347, 351) as having been identified among the plant remains from Peru. This species is cultivated at high altitudes from Mexico to Peru, and its greatest diversity is found in the Andes. The plant is characterized by an over-all pubescence, and the purple, partly hooded flowers have a distinctive folding between the lobes. The curved seeds are dark and usually wrinkled. The pods may be three times as long as they are broad, while others may be one and one-half times as broad as they are long.

Capsicum sinense *Jacques* Hort. Vind. 3 (1776) 38. t. 67.

Capsicum sinense appears to be the most widely distributed of the cultivated peppers in South America at the present time (Smith and Heiser, 1957). However, its place of origin has yet to be determined. The fruits, varying in shape from spherical to elongate, are from one to twelve cm. long. The orange, yellow, red or brown pods have a smooth but sometimes wrinkled surface. A characteristic of the calyx usually present, which differentiates this species from other cultivated peppers, is a marked constriction at the base. However, this character should not be confused with a wrinkled fold that is sometimes found at the base of the calyx in fruits of *C. annuum* that is caused by the enlargement of the fruit.

Heiser (unpublished report) has identified certain of the archaeological remains of *Capsicum* recovered from Huaca Prieta as *C. sinense*; it is probable that some of the *Capsicum* specimens from other sites will be assigned to this species upon a reappraisal of the material (pl. IV, 7).

Capsicum sp.

Some pepper pods found at the pre-ceramic site of Huaca Prieta (Bennett and Bird, 1949, p. 120) and pods, both broken and intact, and stems with and without the calyx recovered in Inca and Inca-associated levels at Pachacamac (Towle, 1948) (pl. IV, 7) have not been completely identified. It is interesting that no peppers were reported among the plant remains found at Ancón by Reiss and Steübel (Wittmack, 1888, p. 346). *Capsicum* was represented at Huaca del Loro and Estaquería in the Nazca Valley (Towle, 1956). According to Wiener (1880, p. 620), the fruits of *aji* were used as models for clay reproductions, and Safford (1917a, p. 22) mentions a clay vessel decorated with two reproductions of the fruit of a *Capsicum* (pl. XIV).

Cyphomandra *Mart ex Sendtn.*

Cyphomandra is a South American genus consisting of about forty species which occur mainly in Brazil. The plants, commonly known as "tree-tomatoes," are shrubs or small trees with three-lobed or pinnately-cut leaves and small flowers in racemes or clusters. The fruits are two-celled, many-seeded berries. Safford (1917a, p. 21) states that certain fruits reproduced in prehistoric Peruvian pottery apparently belong to the genus *Cyphomandra.* These fruits are smooth-skinned and striped like those of *Solanum muricatum*, but they are more pointed and egg-shaped. Safford mentions the possibility of two species being represented: *Cyphomandra betacea*, a widely cultivated form with short, obtuse calyx-lobes, and *C. calycina*, with the calyx-lobes long and acute. He further identifies (*ibid.*, fig. 5) the two fruits constituting a vase recovered at Trujillo as those of *C. calycina.* However, this species is a native of Brazil. Yacovleff and Herrera (1934, p. 275) mention the similarity between the fruits of *Solanum muricatum* and still another species of *Cyphomandra*, *C. splendens*. This species is a native of Peru and is grown in the *montaña* region and the temperate valleys of the Andes for its edible fruits, which are used in soups and relishes. Both this plant and *Solanum muricatum* are known by the vernacular name of *pepino*. The fruit of *C. splendens* is also known as *pimiento* and *tomate* in Cuzco, according to Herrera (1941b, p. 368).

NICOTIANA *L.*

Approximately fifty species, chiefly South American, constitute this genus. These plants are mostly herbaceous annuals and perennials with unbranching stems, which in some species reach a height of six to nine meters. The stems and simple, alternate leaves, often covered with glandular hairs, contain the alkaloid, *nicotine*. The long, tubular flowers, white, greenish, purple or yellow in color, are borne in terminal panicles or racemes and are strongly fragrant. A few species are grown as ornamental plants, but the importance of the genus rests upon two cultivated Nicotianas of economic value: *N. tabacum* L., the source of commercial tobacco, and *N. rustica* L., from which nicotine sulphate, an ingredient of insecticides, is obtained. These two species were also widely used by the aborigines in the New World at the time of Columbus. Both plants are hybrids of species native to the mountainous regions of South America. *N. tabacum* is believed to have originated in the eastern valleys of the Andes, and its use as a masticatory apparently spread among the people of the Amazon basin as far as the north coast of South America and thence to the West Indies. *N. rustica* is a hybrid whose center of origin has been placed in the area about the borders of Peru and Ecuador (Sauer, Carl O., 1950, p. 523). Its use was adopted by the tribes of Central America, Mexico and North America as far north as Canada. It was used as a masticatory in the southern section of its area of distribution, namely Mexico and Central America, but in North America the primary means of consumption was by smoking.

A number of native species of *Nicotiana* are reported from Peru. Several of these were mentioned by the early chroniclers as having been employed by the pre-Columbian Peruvians as "medicine," and their use is said to continue at the present time (Yacovleff and Herrera, 1935, pp. 44–45; Wittmack, 1888, p. 345). There have been no actual, archaeological specimens of the plant recovered from Peruvian sites nor have any representations of it been identified. However, its importance in the pre-Columbian economy of the New World and especially its use as a masticatory in the Peruvian Andes before the Conquest makes it worthy of inclusion here. Its origin, use and development as an economic plant pose a question regarding its historical relationship to coca, another important masticatory plant of the region.

Solanum *L.*

The genus *Solanum* consists of approximately two thousand species native to the tropical and subtropical regions of both hemispheres, but chiefly America. These plants are both wild and cultivated and range from herbs to small trees. They are often covered with spines and thorns. About 150 species are tuber-bearing and to this group belongs the well-known cultigen, the potato.

Solanum muricatum *Ait.* Hort. Kew 1 (1789) 250.
Solanum variegatum Ruíz et Pav. Fl. Peruv. 2 (1794) 32.
'*Kachan,* Aymara (Mejía Xesspe, 1931, p. 11).–*Kachuma,* Aymara (*ibid.*).–Melon pear (Bailey, 1917, vol. VI, p. 3182).–*Pepino* (*ibid.*).–*Xachun,* Quechua (Mejía Xesspe, 1931, p. 11).

The *pepino* is an erect, bushy shrub, up to one meter in height, with entire oblong-lanceolate leaves of a maximum length of sixteen cm. The rather small, light blue to purple, five-petaled flowers are borne in long-stalked clusters. These produce edible fruits, ovoid or egg-shaped, ten to fifteen cm. long, with a smooth yellowish skin marked by longitudinal but discontinuous streaks of violet-purple. The *pepino,* a native of Peru, is widely grown as an herbaceous perennial on the Peruvian coast as well as in other temperate areas of the American tropics. The cultivated forms are propagated by cuttings of the growing shoots, since only vestigial seeds are produced. The early chroniclers mention the growing of *pepinos* on the coast in pre-historic times. However, as would be expected, there is no reference to the remains of these soft, pulpy fruits having been recovered from archaeological sites. Nevertheless, pottery representations of the *pepino,* molded in pairs or groups of several fruits, are fairly frequent (pl. XIII, A). These all bear the characteristic, longitudinal markings. Yacovleff and Herrera, (1934, fig. 1, e, p. 273) illustrate a stirrup vase found at Paracas which consists of two parallel, molded *Solanum variegatum* fruits with joining stems. Reproductions of this fruit are also among the pottery vessels of the Salinar Period on the north coast (Larco Hoyle, 1946, p. 156). A jar representing two fruits was recovered from Chimbote (Safford, 1917a, fig. 4, p. 20), while a single fruit constitutes the body of still another vessel of Mochica ware (Larco Hoyle, 1938, vol. I, p. 78).

Solanum spp.
Potato.–*Papa,* Quechua (Hawkes, 1947, p. 211).

Wild species of tuber-bearing Solanums are found from Colorado to southern Chile, and at the time of the Conquest several were already under cultivation in the Andean region of South America. In fact, the area of high altitudes in Peru and Bolivia is thought by some authorities to be the center of origin of the cultivated potato since it is here that the greatest concentration of species is encountered.

The cultivation complex of the tuber-bearing Solanums consists of a series of species, with a basic chromosome number of twelve, that range from diploid through triploid and tetraploid to one known pentaploid form. Representatives of these groups are grown today in varying degrees from Colombia to Bolivia, as well as on the temperate coastal plain of Chile. The Solanums cultivated in the lower altitudinal limits are for the most part more closely related to the wild forms than those grown in the higher reaches of the Andean highlands.

Among the species cultivated in the latter region are several of the diploid group, the most notable being *S. stenotomum*, commonly known as *chapina*. This plant produces tubers with black flesh which are used as a source of dye rather than for food.

At the higher altitudes, above 12,000 feet, which marks the altitudinal limit of maize, certain triploid species of potato, because of their frost-resisting quality, constitute the dominant cultivated forms. However, their tubers are bitter and must be treated in some way before they are available as food. The Indians of the highlands have long been aware of this characteristic, and they divide potatoes into two groups: the *chchoqhe* which are edible after ordinary cooking and the *lukki* which must be specially prepared by a process that was evolved during prehistoric times. Two products, *chuño* and *moray* (*tuna* or "white" *chuño*) are usually made from the bitter species, although non-bitter tubers are sometimes used. *Chuño* is a basic ingredient of *chupa*, the native stew of the highlands, and a flour is made from *moray*. The method of preparation is similar during the first stages, which consist of freezing, thawing and crushing. At this point, those tubers to be used as *chuño* are dried before storing, while those to be made into *moray* are placed in water and soaked for a period of two months, after which they are dried in the sun. In both cases the tubers, although often much reduced in size, retain their original shape and external characters, but the *moray* tubers are white and lighter in weight than the *chuño*.

The most widely distributed and best known forms of the cultivated potato, however, are the tetraploid Solanums. Hawkes (1947), who has made extensive studies of the potato, considers this group as one large species complex which he designates as *Solanum tuberosum*, with *S. andigenum* of the northern Andes from Peru and Bolivia and *S. chileanum* of the southern Andes and the cool coastal plain of Chile as geographical subspecies. This is in disagreement with the Russian botanists (Bukosov, 1933) who recognize *S. andigenum* as the major species with the widest distribution and *S. chileanum* as a distinct species of the Chilean or southern region. Still other botanists (Correll, 1952) interpret *S. tuberosum* and *S. andigenum* as ecological forms of one species. This confusion in nomenclature is reflected in the archaeological literature of the Andean region, some authors using *S. tuberosum* while others employ *S. andigenum.*

Solanum andigenum is the most commonly cultivated potato in the Andes, and it shows great variation in the color, taste and shape of the tubers. Some ninety varieties have been distinguished. These are cultivated at the present time by the Aymara farmers of the highlands, who have, over the centuries, developed a remarkably extensive vocabulary covering the various types and characteristics of this major starch crop (LaBarre, 1947).

Solanum curtilobum, the only pentaploid species, is grown locally in Peru from Cuzco to Potosí and has a bitter tuber used only for *chuño.*

The history of the potato as a cultivated plant in the Central Andean region is largely derived from archaeological data. In point of time, the earliest specimens of actual tubers that have been found were recovered from the ruins of a house at Chiripa on the shore of Lake Titicaca. This site has been assigned to the Formative Epoch and the charred vegetal remains found there constitute one of the few collections of archaeological plant remains from the Andean highlands. Twelve of the specimens, all of which are carbonized, are in all probability tubers, and since root crops were a staple source of food in the region, it seems possible that one or more of the cultivated genera may be represented. Several of the specimens closely resemble potatoes which may have been stored as raw tubers or as *chuño* (Towle, 1957).

All other evidence relevant to the history of the potato is derived from the coast. Here, the earliest manifestations are representations of the tubers and stylized motifs derived from them which occur in Mochica ceramics. This use of the potato in pottery design continued on the north coast into the Chimu and later Inca periods (pl. X, B; XI, A, B; XV, B). The illustrations of such material in Salaman (1949) adequately show the various ways in which the subject was treated, from simple, naturalistic representations to highly stylized forms heavily burdened with religious and often enigmatic symbolism. Harms (1922, p. 184) describes three vessels, one realistically depicting "a demoniac face," the other two, handled jars of black ware. From the description, these seem to have been also derived from the north coast, although Harms does not mention the locality from which they came. Three pottery vases from Chimbote are shown by Safford (1926, pl. 7). One of these consists of two tubers of natural clay color, the other two, like those referred to by Harms, are of black ware. One of the latter is a whistling jar. All have prominent "eyes," a characteristic often emphasized.

Although Nazca pottery, like that of Mochica, was a well-developed art and tells us much of the everyday life of the people of the south coast and the plants known and used by them, representations of the potato are lacking. In fact, from the available evidence, it was not until the later Fusion periods that we find the potato as an art motif in this area. Yacovleff and Herrera (1934, p. 297) mention a unique representation in the National Museum at Lima, which came from the site of Pacheco.

Coastal specimens of actual tubers have been found only on the central coast at sites assigned to the Fusion and later Periods. This would be expected, as by that time there was much cultural interchange between the highlands and the coastal area. Although Wittmack (1888, p. 340) found no specimens of the potato at Ancón, Uhle recovered a number during his extensive explorations at Pachacamac. Harshberger (1898, p. 3) describes these as for the most part small, about one inch in diameter, and resembling the tubers of the wild potatoes which he had collected in Mexico. One small, well-preserved, slightly compressed specimen, also from this site, was among the remains from the lower Inca levels examined by Towle (Towle, 1948) (pl. I, 2). It was not identified as to species. Safford (1926) states that specimens of "chunyos" were among the Peruvian material in the American Museum of Natural History. These came from pre-Columbian sites and are included among the articles of food pictures on plate 6, no. 2 (*ibid.*). Unfortunately, the sites from which these came is not given.

It is indeed interesting that the knowledge of this important plant of the highlands should apparently be confined during the early periods to the north coast. Salaman (1949), who has spent many years in the study of the potato and its history, has suggested that the potato, both as a raw tuber and as *chuño*, was transported as an item of trade from the highlands into the north coast valleys. Although maize was the basic starch crop of the coast, the potato with its accompanying lore and religious connotations was soon adopted by the coastal peoples. Here would seem to be further evidence of highland-coastal diffusion. Although certain wild species of tuber-bearing Solanums grow on the Peruvian coast today, Salaman thinks it doubtful that the potato was actually cultivated in the coastal valleys in prehistoric times because of the climatic conditions existing there.

BIGNONIACEAE. Trumpet-Creeper Family.

ARRABIDAEA *DC.*

The forty-five species of this genus are native to the tropical and semi-tropical regions of America.

Arrabidaea Chica *Verl.* in Rev. Hortic. (1868) 154.
Bignonia Chica Humb. & Bonpl. Pl. Aequin. (1808) 107, t. 31.

This climbing vine is widely distributed in the tropical areas of South America. Its large purple flowers produce linear pods up to sixty cm. in length. The leaves are the source of a pigment, the color of which varies from orange and red to purple depending upon the method of preparation. The pigment is used by the Indians for a body paint, as a dye for fibers and basketry materials and for painting pottery. It was formerly exported under the name of *chica* or *carajura* and was used in Europe for dyeing textiles. Rochebrune

(1879, pp. 347, 355) reports finding the fruit of *Bignonia Chica* among the plant remains of coastal Peru.

CRESCENTIA *L.*

The five species of evergreen trees comprising this genus are native to tropical America. *Crescentia Cujete* is cultivated in both the Old and New Worlds for its gourd-like fruits.

Crescentia Cujete *L.* Sp. Pl. (1753) 626.
Calabash-tree.—*Cujete,* Mexican name (Arbelaez, 1956, p. 220).— *Pamuco* (Yacovleff and Herrera, 1934, p. 315).—*Tutuma* (*ibid.*).

The calabash-tree, five to eight meters high, has a short trunk and long spreading branches. The lanceolate leaves, up to fifteen cm. long, are nearly sessile and often grow in clusters. The yellowish-purple to green tubular flowers are borne close to the trunk and produce nearly globular fruits, often thirty cm. in diameter. For centuries, these gourd-like fruits with their thin, woody, durable shells, have been widely used as containers. In some cultures religious or medicinal properties are attributed to the soft, white pulp in which the seeds are embedded. This tree is reported from northeastern Peru and the tropical inter-Andean valleys. Although it appears not to have been native to the coast, several specimens of *Crescentia* fruits have been found in archaeological coastal sites. Two calabashes, one oval, the other elliptical, but both retaining the natural green-yellow color and having an aperture cut in one end, were recovered from Paracas (Yacovleff and Herrera, 1934, p. 316). An oblong fruit filled with the hardened resin of *Myroxylon* sp. came from a grave at Ancón (Safford, 1917a, p. 22, fig. 9). A spoon or ladle recovered from a grave at the Gallinazo site of Huaca de la Cruz in the Virú Valley was identified by Junius Bird as made from the fruit of a *Crescentia,* "perhaps *Cujete L.*" (Towle, 1952c, p. 356). Wiener (1880, p. 601) states that pottery vessels were made by the pre-Columbian Peruvians to resemble the fruits of the calabash-tree.

ACANTHACEAE. Acanthus Family.

DICLIPTERA *Juss.*

A genus of branched, perennial herbs consisting of about eighty pantropical species.

Dicliptera Hookeriana *Ness.* in DC. Prodr. 11 (1847) 480.
Dicliptera peruviana *Juss.* in Ann. Mus. Par. 9 (1807) 268.

The leaves of these closely related species, native to Peru, were reported by Rochebrune (1879, pp. 347, 355) to have furnished a dye which was used by pre-Columbian Peruvians. He adds that when this was mixed with the blue and yellow dyes obtained from *Piper lineatum* and *Lafoensia acuminata*, a green color was obtained.

RUBIACEAE. Madder Family.

Relbunium *Benth. and Hook.*

Relbunium comprises about twenty-five species of annual and perennial herbs native to the tropical and temperate regions of America from Mexico to Argentina. Several of these species are mentioned in the literature as a source of red dye, both in pre-Columbian and in modern times (Yacovleff and Herrera, 1935, pp. 75–76; Fester, 1953). This genus is closely related to *Rubia* of the Old World, which furnished madder dyes.

Relbunium nitidum (*HBK*) *Schum.* in Mart. Fl. Bras. 6, pt. 6 (1888) 114.
Rubia nitida HBK, Nov. Gen. et Sp. 3 (1819) 339 (*Relbunium* sp.).

This species occurs in the *lomas* region of Peru (Weberbauer, 1945, p. 257). Rochebrune (1879, pp. 347, 355) reports the occurrence of leaves and roots of *Rubia nitida* at Ancón, adding that a red dye was obtained from the roots. Fester (1953) believes that the red dye used in the Paracas textiles was very probably obtained from the roots of *Relbunium nitidum.*

CUCURBITACEAE. Gourd Family.

Cucurbita *L.*

The twenty species which comprise this genus are native to the warm and cool temperate regions of the New World.

Cucurbita ficifolia *Bouche* in Verh. der Ver. des Gartenb. Berlin, 12 (1837) 205.

Cucurbita ficifolia is a long-running, perennial vine with abundant serrate and moderately lobed leaves, twelve to eighteen cm. across. The bright yellow to light orange flowers produce fruits of variable sizes. These are commonly oblong to nearly globular in shape, with white flesh and a firm rind, usually green marked with white stripes and blotches. The mature peduncle is hard and is either not enlarged or only slightly expanded at the point of attachment to the fruit. The seeds are black, brownish black or a dingy white in color and average two cm. in length. This species, usually cultivated as an annual, tolerates cool temperate climates and is grown from Mexico to Chile, its area of distribution exceeding that of all of the other cultivated cucurbits (Bukosov, 1930).

Although the place of origin of *Cucurbita ficifolia* has not yet been determined, its present distribution points to the temperate highlands of Middle or South America. However, evidence of the use of this cultigen in pre-Columbian times has been found at only one archaeological site. At Huaca Prieta, on the northern coast of Peru, black seeds of *C. ficifolia*, along with the fringed-margined seeds of *C. moschata*, were recovered from pre-ceramic levels (Whitaker and Bird, 1949). In addition, a dingy-white-seeded variety was represented in still later levels assigned to the Cupisnique Period (*ibid.*). Although this species, widely cultivated in the Andean region at the present time, was one of the earliest food crops of the Americas of which we have record, it has not been reported from archaeological sites in Peru later than the Cupisnique levels at Huaca Prieta.

Cucurbita maxima *Duch* Lam. Encyc. 2 (1786) 151.
Pepo maxima Peterm. Fl. Lips. Excurs. (1838) 562.
Sapallu, Quecha (Mejía Xesspe, 1931, p. 11).—*Zapallo,* Spanish for the above. A term commonly applied to the fruits of cucurbits (Yacovleff and Herrera, 1934, p. 303).

C. maxima is an annual vine, generally long-running, with serrate, sometimes lobed, orbicular or reniform leaves and large, spreading, light to dark yellow flowers. The fruits, often very large, are spherical, oblong or turban-shaped with more or less spongy, short, cylindrical peduncles which do not expand at their attachment to the fruit. The ovate-ellipsoidal seeds are clear white or fawn-colored, with the apex ordinarily oblique and the base rounded. The seed margin is smooth and usually similar to the body of the seed in texture and color, although in some instances the margins may be slightly paler. To this species belong the autumn and winter squashes, such as the mammoth, Hubbard and turban varieties.

Cucurbita maxima is a native of South America, and since its nearest relatives are found in Uruguay and Argentina, its place of origin probably was east of the Andes. This squash is not known to have reached the Caribbean coast of Colombia before the arrival of the Spanish. However, it was a widely used cultigen at that time in both the warm and temperate areas of the Inca Empire, and we have archaeological evidence of its use on the central and southern Peruvian coasts at even earlier dates (pl. IV, 4). Among the reproductions of vegetable foods in Chimu and Mochica pottery, there are a number of excellent specimens of various squash forms. Safford mentions one of these as an exact reproduction of a warty Hubbard squash (1917a, p. 18). Wittmack and Naudin, an accredited botanist of his day, identified seeds of *C. maxima* among the plant remains from Ancón (Wittmack, 1880–1887, pl. 107, fig. 15). Although Ames (1939, p. 80) questions their identification on the grounds that the material was not adequate, Harms (1922, p. 188) accepts their determination and adds that he found seeds of this plant in a container from Pachacamac. These specimens were rounded, 2.5–3 cm. in length and 1.3–1.6 cm. in breadth, and were yellowish-brown to a dingy white color. Carter reports identifying four seed fragments of *C. maxima* recovered from San Nicolás near Supe, a site placed in the Supe Middle Period about 1200 A.D. (Carter, 1945). This author also mentions finding 108 seeds of this species at Ocucaje in the Ica Valley on the south coast, a site assigned to the Proto-Nazca Period, circa 600 A.D. He further mentions 224 seeds from Chulpaca, another site in this valley, but one belonging to the later Middle Ica II Period, 1300–1400 A.D. (*ibid.*).

Cucurbita moschata *Duch.* in Poiret in Ann. Sci. Nat. 11 (1818) 234.

Crookneck or warty squash.—*Calabaza,* Cuzco (Herrera, 1939, p. 130).—*Lacayote,* near Cuzco (Herrera, 1939, p. 217; Yacovleff and Herrera, 1934, p. 304).—*Zappalo,* a term generally applied to fruits of the cucurbits (Yacovleff and Herrera, 1934, p. 303).

Cucurbita moschata is an annual vine with large, broad-ovate to nearly orbicular-ovate leaves which are occasionally lobed and often marked with whitish blotches. The yellow to yellow-orange flowers develop fruits of many shapes. The most common forms are oblong or crook-necked, and these contain a dry to moist flesh, yellow to orange in color. The mature peduncle, frequently long and slender, is hard and five-angled and becomes enlarged at its attachment to the fruit. The seeds are plump and flat, ovate to elliptical in outline, with a slanting seed-scar. The margin of the seed, which is often fringed, extends around the entire edge and is darker in color and different in texture from the body of the seed.

The crookneck squash is apparently of Mexican or Central American origin, and is extensively cultivated at the present time in these areas, as well as in Colombia and some of the islands of the West Indies. According to the archaeological evidence, *C. moschata* was the most widely distributed of the cultivated species of the genus in pre-Columbian America. Specimens have been recovered from archaeological sites in the southwestern United States, Central America and Peru. However, Whitaker and Bohn (1950, p. 67) believe that the archaeological material of *C. moschata* should be reexamined, since it is possible that some of the specimens assigned to this species may belong to *C. mixta.*

Seeds of *C. moschata* were found in pre-ceramic levels at Huaca Prieta and the warty shells and seeds of a variety of this species were recovered from a later level in association with maize and Cupisnique pottery (Whitaker and Bird, 1949, p. 14). Whole and broken peduncles of this squash have been identified from the Gallinazo horizon at Castillo de Tomaval (pl. IV, 3) and Huaca de la Cruz in the Virú Valley (Towle, 1952c, p. 355). Wittmack (1880–1887, pl. 107, fig. 16) illustrates a seed of *C. moschata* which was recovered from Ancón, the identification of which was confirmed by Naudin. A number of seeds were found in a container at Chuquitanta (Harms, 1923, p. 186), and the lower end of a crookneck squash with seeds still embedded in the dry flesh was recovered from the Inca-associated level of Pachacamac (Towle, 1948) (pl. IV, 3). At Vista Alegre, seeds were found in Decadent Maranga levels, and a number were included with other plant remains in a textile bag from a grave of undetermined cultural identification (Towle, 1958). Carter (1945, p. 165) reports that numerous seeds of *Cucurbita moschata* were found at two sites of Chincha. A few seeds or shell fragments of this species were recovered at Cahuachi, Huaca del Loro and Estaquería in the Nazca Valley on the southern coast of Peru (Towle, 1956). Various forms of the fruit were reproduced in Mochica pottery (Larco Hoyle, 1938, vol. I, fig. 36; Yacovleff and Herrera, 1934, fig. 26) (pl. VI, A, B). Yacovleff and Herrera further state (*ibid.*, p. 303) that containers of a yellow-white ware, which appear to have been made in imitation of different forms of the crookneck squash, have come from Paracas.

CYCLANTHERA *Schrad.*

A genus of about forty species native to the Americas.

Cyclanthera pedata (*L.*) *Schrad.* Ind. Sem. Hort. Goett. (1831).
Acho'qcha, Quechua (Mejía Xesspe, 1931, p. 11).–*Acha'qcho,* Aymara (*ibid.*). –*Caygua* [*Caigua*] (Yacovleff and Herrera, 1935, p. 46).

Caigua is cultivated for its edible fruit on the Peruvian coast and in the temperate Andean valleys, its natural distribution extending from Bolivia to Mexico. The plant, like some of the related species of *Cucurbita,* is a trailing vine with digital leaves of five to seven leaflets. The flowers are arranged in loose panicles. The fruits may have either a smooth skin or one characterized by blunted prickles or beaks protruding from the surface. According to Herrera, the type cultivated in the region of Cuzco produces a large smooth fruit (Macbride, 1937, pt. 6, no. 2, p. 372).

The finding of molded representations of *caigua* among the Cupisnique ceramics on the north Peruvian coast, as reported by Larco (1946, p. 150), indicates the knowledge and possible uses of this species as a food plant during the Formative Epoch.

LAGENARIA *Ser.*

A monotypic genus occurring in both the Old and New Worlds.

Lagenaria siceraria (*Mol.*) *Standl.* Field Mus. Publ. Bot. 3 (1930) 435.
Cucurbita lagenaria L. Sp. Pl. (1753) 1010.
Cucurbita siceraria Mol. Sagg. Chile. (1782) 133.
Lagenaria leucantha (Duch.) Rusby in Mem. Torrey Bot. Club 6, no. 1 (1896) 43.
Lagenaria vulgaris Ser. in Mem. Soc. Phys. Genev. 3 (1825) 25.

Gourd.—*Calabazo* (Yacovleff and Herrera, 1934, p. 312).—*Matti*, Quechua for the globular form of gourd; the word from which the commonly used *maté* is derived (*ibid.*, pp. 312–313).—*Puru*, Quechua for the bottle-shaped fruit of *Lagenaria*; sometimes spelled *poro* (*ibid.*).

Lagenaria siceraria is a white-flowered, trailing, annual vine up to seven meters in length, with cordate-ovate or broader, unlobed leaves. The hard-shelled fruits, from 7.5 cm. to nine dm. in length, are of varying shapes and sizes, the most common of which are the Hercules club, bottle-, dipper-, trough-, dumb-bell, and calabash-gourds. However, this latter term is unsatisfactory because of the resulting confusion with the fruit of the calabash tree (*Crescentia Cujete*). The seeds of *Lagenaria* are tan or grayish in color and differ in shape and general appearance depending upon the variety of fruit from which they come. They are more or less tapering, oblong or truncate at the base, with parallel, longitudinal ridges along the inner edge of the seed-margin. Paired, winged protuberances at the base of the seed opposite the hilum are usually present. *Lagenaria* is widely cultivated now, as it was in prehistoric times, for its durable, mature fruits which are easily converted into containers of various sorts. Although probably a native of the Old World, this plant was cultivated in the western hemisphere before 1492, as is attested by the finding of remains of the fruit as well as gourd seeds in archaeological sites in North and South America. However, there are no historical references to the young fruits having been used as food, as they are in certain parts of the Old World.

Fruits of this cucurbit are of common occurrence in the archaeological sites of coastal Peru. They served the prehistoric Peruvians not only as containers but also as models for pottery vessels (Yacovleff and Herrera, 1934, p. 314). The custom of decorating gourds by pyrogravure, carving and lacquering, was developed at an early date, and the archaeological evidence of fruits of *Lagenaria* having been worked covers a wide range of time and space. At Huaca Prieta on the north coast of Peru, abundant proof of use of the gourd was found in pre-ceramic and Cupisnique levels (Whitaker and Bird, 1949). From the numerous fragments, some few of which were decorated, it was determined that containers, ladles or scoops and discs were well represented. Also eight gourds, still attached to a fish-net, were recovered. In addition to the fruits, two types of *Lagenaria* seeds were isolated: a large, broad type from the pre-ceramic horizon, and a smaller type from both pre-ceramic and Cupisnique levels (*ibid.*; Towle, 1952a). Also from the north coast in the Virú Valley, a fragment of the stem-end of a gourd and twelve pieces of a fruit-shell were recovered from the pre-ceramic level of the excavation at Huaca Negra. At Castillo de Tomaval, which has been placed in the Gallinazo Period, a single seed and numerous shell fragments were found, and seeds and pieces of the fruit of *Lagenaria* were identified from the plant remains recovered from the

Gallinazo and Mochica levels of Huaca de la Cruz (Towle, 1952c, pp. 355–56) (pl. IV, 2).

Numerous specimens of gourds, both whole and broken, have been reported from the central Peruvian coast. Specimens of gourd utensils were recovered from Ancón (Reiss and Stübel, 1880–1887, vol. 3, pl. 82, figs. 1–13), as well as a seed identified by Wittmack as that of *Lagenaria* (1880–1887, pl. 107, fig. 17). Harms noted parts of the shell of a gourd and seeds in the mummy wrappings from a burial found at this site (Harms, 1922, p. 186). Both whole and broken gourds, some of which were decorated, and seeds are described by Whitaker from Ancón (1948, pp. 57, 62); these are assigned to the Middle Periods from 900–1200 A.D. This author further describes containers, bottles and cylindrical tubes made from the stem-end of gourds from Chancay; these he places in the Late Periods, 1300–1400 A.D. (*ibid.*, pp. 57–58, 62). A whole gourd with seeds, two bottles and two bowls made from this fruit are described by Towle from the Aspero site near Puerto de Supe (Towle, 1954, pp. 136–37) (pl. IV, 2). The remains of a small gourd dish from the Lighthouse site in the same vicinity are also described (*ibid.*). Two bowls and a fruit containing seeds were found in a grave in the Chillón Valley (Yacovleff and Herrera, 1934, pp. 314–15).

Still another site on the central coast has furnished us with large numbers of gourd specimens. From Pachacamac Towle identifies numerous pieces of broken shells and seeds recovered from Inca-associated and Inca levels (Towle, 1948). From the Berthon collection of plant remains from excavations of sites in this area of the coast, Costantin and Bois (1910, p. 259) identify containers made of these fruits as well as a seed, which they illustrate.

The use of *Lagenaria* was also prevalent on the south coast. From Chincha, gourd vessels of different shapes, both plain and decorated, and unworked specimens containing seeds are described by Whitaker as belonging to the Late Chincha and Inca Periods, 1300–1500 A.D. (Whitaker, 1948, pp. 54–55, 62).

Carter also identifies seeds from Chincha (Carter, 1945, p. 166). Sites in the Ica Valley have contributed additional specimens recovered from the Ica, Middle Ica, Late Ica, Inca, Late Inca and Epigonial Periods (Whitaker, 1948, pp. 55–57, 62). Yacovleff and Herrera (1934, p. 315) mention three gourd containers and a tube from the stem-end of a gourd which were recovered from Paracas Cavernas (dated by Whitaker [1948, p. 62] as belonging to the Chavín Periods of the south coast, 400–600 A.D.). Towle describes a small bottle and a shallow dish which had been enclosed in the wrappings of a mummy found at Paracas Necropolis (Towle, 1952b, pp. 239–40); this site is assigned to the Early Periods of the south coast, 700–800 A.D. (Whitaker, *op. cit.*). Specimens of broken gourds were of common occurrence at Cahuachi, Huaca del Loro and Estaquería in the Nazca Valley (Towle, 1956). Two whole fruits of *Lagenaria* and gourd bowls are reported to have been recovered from Nazca by Yacovleff and Herrera (1934, p. 315), and Nazca pottery made to resemble the fruits of *Lagenaria* is represented by these authors.

Thus we have archaeological evidence of this ancient cultigen having been generally distributed at various times from the Chicama Valley to the Rio Grande de Nazca, and our knowledge of its use as an economic plant on the coast of Peru extends from the time of the Early Farmer on the north coast to that of the Late Inca Period on the central and south coasts.

COMPOSITAE. Sunflower Family.

BACCHARIS *L.*

Baccharis is an American genus of about three hundred species of herbs and shrubs with alternate leaves and clusters or panicles of small white or yellowish flowers. In Peru it is one of the most frequently encountered genera of the *Compositae.*

Baccharis polyantha *HBK.* Nov. Gen. et Sp. 4 (1820) 64.
Baccharis prostrata *Pers.* Syn. 2 (1807) 424.
Chchillca (Yacovleff and Herrera, 1935, p. 45); Cuzco (Herrera, 1939, p. 167).– *Chilca* (Yacovleff and Herrera, 1935, p. 45); for *B. polyantha* (Saffray, 1876, p. 401).

The Peruvians before the Spanish conquest (Yacovleff and Herrera, 1935, p. 45), as well as those Indians now living in the Cordillera (Saffray, 1876), obtained a green or yellow dye from these species of *Baccharis.*

Baccharis sp.

Spindles made from the stems of a *Baccharis* that still retained their bark were recovered at Paracas (Yacovleff and Herrera, 1935, p. 45).

MUTISIA *L.*

A South American genus of about sixty species.

Mutisia hirsuta *Meyen,* Reise (1834), 451.
Mutisia vicaefolia Cav. var. *hirsuta* Meyen.
Chinchircuma, Aymara (Yacovleff and Herrera, 1935, p. 81).

This shrub, 1.5 to three meters high, has a slender stem, few branches and many alternate, oblong leaves. The yellow-orange inflorescence is a solitary, terminal head, composed of minute, labiatious disc-flowers and a few ray flowers. The former are attractive to humming-birds because of the nectar they contain. This species is endemic to the Department of Cuzco, where it grows at altitudes of 300 to 7,000 meters. The stems of the plant are burned for the ashes, which are used in the consumption of coca, and the flowers often decorate signs announcing the selling of *chicha* (Herrera, 1923, p. 444). Representations of the flowers of this plant frequently appear on Inca vessels. A decorative floral band that adorned a *kero* is illustrated by Yacovleff and Herrera (1935, fig. 62). These floral elements have been identified by the authors as those of *chinchircuma, ccantu* and *ñucchu.*

Polymnia *L.*

The ten to twelve species that constitute this genus are native to America from Canada to Argentina.

Polymnia sonchifolia *Poepp. et Endl.* Nov. Gen. et Sp. 3 (1845) 47, t. 254.
Polymnia edulis Weddell in Ann. Sci. Nat. ser. 4, t. (1857) 114.
Aricoma (Cardenas, 1948, p. 49).—*Llacjon* (Herrera, 1941b, p. 419).—*Llacon,* Spanish form from the Quechua (Herrera, 1941a, p. 17).—*Llaq'on,* Quechua (*ibid.*).—*Yacon,* Spanish form from the Quechua (*ibid.*).

Yacon is grown as a minor food crop in the temperate valleys of the Andes from Colombia to Argentina. It is an herbaceous plant, one meter or more in height, with large, alternate leaves, the upper surfaces of which are covered with a dense pubescence. The small, yellow, composite flowers resemble those of the "single" dahlia, to which the plant is closely related. It produces edible, spindle-shaped roots, ten to twenty cm. in length and three to four cm. in diameter, which contain sugar and some starch. Their sweetish, agreeable flavor is said to be improved when they are exposed to the sun. The roots are preferably eaten raw, since cooking destroys their crispness and pleasant flavor. The people of pre-Columbian Peru and Bolivia cultivated *yacon,* and in Cuzco, in the colonial period, the roots were eaten especially during the fiesta of Corpus, which represented the ancient ritual of Cojapac-raimi (Cardenas, 1948, p. 49). Safford (1917a, p. 24) states that he found actual specimens of the roots of *yacon* among the Peruvian plant remains at the National Museum in Washington. He also identifies molded reproductions of the roots as part of a prehistoric vase. These were joined in an upright stem on which was placed the head of a monster with feline teeth. O'Neale and Whitaker (1947, p. 320) have identified the elongated, curved, fleshy roots depicted among the designs of an early Nazca embroidery as those of *Polymnia sonchifolia.*

EXCLUDED SPECIES

References to the following species appear in the earlier ethnobotanical literature of Peru, apparently the result of misidentification.

Cucumis sp.

Certain poorly preserved cucurbitaceous seeds are mentioned by Costantin and Bois (1910, p. 263) as possibly those of a species of the genus *Cucumis.* This is an Old World genus not otherwise represented in pre-Columbian Peruvian sites.

Fouquieria formosa *HBK.*

This plant is mentioned as a source of dye in prehistoric Peru (Rochebrune, 1879, p. 347). However, its presence there at that time is doubtful, for it is a native of Mexico and it is not listed in the modern floras of Peru.

Garcinia Mangostana *L.*

Rochebrune (1879, pp. 346, 351) reported that the fruits of this plant, usually cut in half, had been found in string bags recovered from graves. His identification, however, was questioned by Wittmack (1888, p. 341), since the mangosteen is a native of Asia and was brought to the New World as a horticultural importation well after the period of discovery.

Musa paradisiaca *L.*

Leaves of the banana are said to have been found in Peruvian graves (Rochebrune, 1879, pp. 348, 352; Pickering, 1879, p. 663). Rochebrune (1879, p. 352) also speaks of specimens of the seedless, berry-like fruits having been recovered. However, this Old World species was brought to America only after the European discovery. Since the specimens are not available for study, it is not possible to determine their correct identifications.

ENUMERATION OF PLANTS ACCORDING TO THEIR USES

The species discussed in Part I are listed below under the headings of *Food Plants, Industrial Plants* and *Drug Plants.* A comparison of this list with those of Rochebrune (1879) and Wittmack (1888), who used similar categories, will show some discrepancies, attributable in great part to differences in identification.

As mentioned above, the most comprehensive list of plants constituting the ethnobotany of pre-Columbian Peru is that of Yacovleff and Herrera (1934–1935). This work is based on descriptions of the Peruvian plants included in the accounts written by the chroniclers of the colonial period. Only about half the plants cited, however, have been recovered as actual plant remains or have been identified in pre-Columbian art. All these species are discussed in the present paper.

The majority of contemporary writers on the cultural history of the Central Andes mention only the more common cultivated and wild species that were included in the food economy of the early cultures, with the addition of one or two industrial or medicinal plants such as cotton and coca. The thirty-one species listed by Bennett (1946a, pp. 20–21) are quite representative in this respect. Archaeological evidence in the form of plant remains or as art representations is reported for all but one of these species, namely, *cañihua* (*Amaranthus sp.*).

Food Plants

Beverages

Zea Mays

Condiments

Capsicum annuum
Capsicum frutescens
Capsicum pubescens
Capsicum sp.

Fruits

Ananas comosus
Annona Cherimolia
Annona muricata
Annona squamosa
Aristotelia Macqui
Bunchosia armeniaca
Campomanesia lineatifolia
Carica candicans
Carica Papaya
Caryocar amygdaliferum
Solanum muricatum
Cucurbita ficifolia
Cucurbita maxima
Cucurbita moschata
Cyclanthera pedata
Cyphomandra betacea
Cyphomandra splendens
Lagenaria siceraria
Lucuma bifera
Passiflora sp.
Persea americana
Psidium Guajava

Grains

Amaranthus sp.
Chenopodium pallidicaule
Chenopodium Quinoa
Zea Mays

Legumes

Arachis hypogaea
Canavalia sp.
Lupinus mutabilis
Inga Feuillei
Inga Endlicheri
Phaseolus lunatus
Phaseolus vulgaris

Roots and tubers

Arracacia xanthorrhiza
Canna edulis
Canna sp.
Cyperus sp.
Ipomoea Batatas
Manihot esculenta
Oxalis tuberosa
Pachyrrhizus Ahipa
Pachyrrhizus tuberosus
Polymnia sonchifolia
Scirpus sp.
Solanum tuberosum (and allied species)
Tropaeolum tuberosum
Typha angustifolia
Ullucus tuberosus
Xanthosoma sp.

Industrial Plants

Dyes and tannins

Arrabidaea Chica
Baccharis polyantha
Baccharis prostrata
Bixa Orellana
Caesalpinia Paipai
Caesalpinia spinosa
Indigofera suffruticosa
Roupala ferruginea
Dicliptera Hookeriana
Dicliptera peruviana
Lafoensia punicaefolia
Miconia chrysophylla
Piper asperifolium
Piper lineatum
Relbunium nitidum

Exudations

Myroxylon peruiferum
Prosopis chilensis
Myroxylon sp.

Fibers and stems

Acisanthera inundata
Ananas sp.
Baccharis sp.
Bombax Ceiba
Calamagrostis spp.
Canna sp.
Carludovica palmata
Ceiba pentandra
Chusquea scandens
Cyperus sp.
Furcraea andina
Furcraea gigantea
Furcraea occidentalis
Gossypium barbadense
Guadua angustifolia
Gynerium sagittatum
Phragmites communis
Juncus sp.
Scirpus tatora
Scirpus sp.
Sporobolus sp.
Tillandsia Gilliesii
Tillandsia latifolia
Tillandsia maculata
Typha angustifolia
Zea Mays

Fruits used for utensils

Crescentia Cujete
Lagenaria siceraria

Seeds used for beads

Achras Zapota
Aniba Puchury-minor
Erythrina falcata
Erythrina sp.
Mucuna elliptica
Nectandra Pichurim
Nectandra reticulata
Nectandra sp.
Ormosia sp.
Thevetia peruviana

Woods

Buddleia sp.
Godoya obovata
Guadua angustifolia
Guilielma ciliata
Guilielma gasipaes
Guilielma insignis
Gynerium sagittatum
Inga Feuillei
Kageneckia lanceolata
Pavonia paniculata
Pineda incana
Porlieria hygrometra
Prosopis chilensis
Salix chilensis

Miscellaneous

Opuntia exaltata
(combs, pins)
Phaseolus sp.
(games)
Phytelephas macrocarpa
(vegetable ivory)
Sapindus Saponaria
(vegetable soap)

Medicinal and Masticatory Plants

Andira inermis
Aniba Puchury-minor
Erythroxylon Coca
Mucuna elliptica
Nicotiana sp.
Piper asperifolium
Erythroxylon novogranatense

PART II

ETHNOBOTANICAL EVIDENCE IN TERMS OF CULTURAL CHRONOLOGY

In the following pages the ethnobotanical evidence for the prehistoric cultures of the Central Andes, except for the Epoch of Pre-Agriculture which is described for the region as a whole, is presented for each of the coastal and highland cultural-geographical divisions. These data are described in terms of the cultural sequences that are represented in the several areas, thus giving a chronological picture of the use of wild and cultivated plants against a background of the cultures of which they were part. A reasonable amount of general information has also been included for the benefit of the non-archaeological reader.

Epoch of Pre-Agriculture

Only a few archaeological sites attributable to the Epoch of Pre-Agriculture have been found in Peru. This epoch represents the culture of food-gatherers—in the highlands by those people whose economy was based primarily upon hunting, on the coast by those who depended also upon the sea for their subsistence. In the central highlands east of Lima, evidences of hunter-gatherers have been found in a series of rock shelters at Huancayo (Tschopik, 1946). The artifacts recovered include both crudely formed hunting tools and well-made points that might have been attached to some type of projectile. In addition to these locations in the highlands, there are several isolated camps or workshops of nomadic hunters nearer to the coast between the Virú Valley and Pacamayo, which produced large quantities of pressure-flaked objects (Bird, 1948b, p. 27). In northwestern Peru, additional sites have been investigated which may have been inhabited by a hunting people (Brown, 1926).

Although parts of the Peruvian coast, like that of northern Chile, furnished satisfactory locations for fishing communities, it is only recently that a preceramic, pre-agricultural site comparable to that found by Bird in northern Chile has been discovered (Strong, 1954, 1957). Several low shell-mounds located at the Bay of San Nicolás on the Peruvian South Coast have been assigned to this early epoch. Here a fisherman-hunter people relied for their subsistence upon the abundant marine life of the Pacific, as the remains of shell-fish, fish and sea lions demonstrate (*ibid.*). Although both maize and gourd specimens were found at the San Nicolás mounds, such a cultural sequence as was demonstrated at the sites in northern Chile appears to be lacking. The remains of these two cultigens were recovered only from the mound surface, a fact that makes one hesitate in placing the upper levels of the site in the Epoch of Incipient Agriculture. This caution seems further justified since the type of maize found resembles that associated with much later cultural levels (Towle, 1956).

The subsequent history of the highland hunters is a matter of conjecture. It has been noted that the type of stone work found in these sites is the result of pressure-flaking, a technique of manufacture in contrast to that used in the later Epoch of Incipient Agriculture (Bird, 1948b, p. 27). On the basis of the available evidence, Bird (Bennett and Bird, 1949, p. 117) has pointed out that it is doubtful that these hunter-gatherers were the authors of the cultures later to be found either on the coast or in the highlands.

However, the highland manifestations of the hunter-gatherers show some similarities to those on the coast of southern Peru and northern Chile. Bird (1943; 1946, p. 587), during his careful researches on the middens of Arica, Pisagua and Taltal on the northern Chilean coast, recognized in the lower levels two distinct cultural phases which he designates as the First and the Second Pre-Agricultural Periods. The former he characterizes as having shell fish-hooks, composite fish-hooks, harpoons fashioned for the spearing of sea mammals, as well as coarse, percussion-flaked tools. In the later level, fish-hooks made of thorns replace those of shell found in the preceding period and he finds bone points for fishing or sealing harpoons and chipped stone artifacts, including one type of point that might have been used to tip arrow shafts. Among the specimens recovered at this depth were sticks that might have been used as throwing sticks. These and the thorn fish-hooks are the first such plant remains to be recorded from sites of the food-gatherers, for although the descriptions of pre-agricultural sites in Peru and northern Chile contain lists of the artifacts recovered, they do not include either worked or unworked plant remains.

This lack of vegetal material does not bar the possibility of finding such cultural evidence in dry rock shelters or caves. Among people practicing a food-gathering economy, the major food supply, whether procured by hunting, fishing or gathering, is usually augmented in varying degrees by other food sources that the environment may offer. Specifically, in the case of the botanical environment, not only would the plants that had been tested and found satisfactory be used as food but other species would find their way into other aspects of the culture. Thus in those localities where conditions were favorable to the preservation of perishable materials, plant remains of food and artifacts of plant origin would be expected to be among the specimens recovered.

THE NORTH COAST

It has been suggested that this coastal division, from the Piura to the Casma Valley, be further subdivided into the Far North Coast and the North Coast, the Piura, Lambayeque and Jequetepeque valleys constituting the former, the Chicama, Moche, Virú, Santa, Nepeña and Casma the latter (Bennett, 1948; Willey, Chart 1954). The available archaeological data appears to warrant such a division. Of the Far North Coast valleys, those of the Lambayeque and Jequetepeque are the best known, although some excavating has been done in the others. A number of large sites are located in the Lambayeque Valley, where both Bennett and Kroeber have conducted research. The sites of Batan Grande and Chongoype have produced archaeological materials showing definite Chavín influence, without, however, any evidence of the later Coastal Tiahuanaco. The valley of Jequetepeque contains the Chavín site of Kuntar Wasi. The reports of the excavations of sites in these two valleys do not include descriptions of plant materials. In fact, our information on the ethnobotany of the Far North Coast is restricted to scattered references to pottery representations of fruits, such as a vessel depicting the lucuma from Lambayeque, and the occasional mention of specimens of plant origin and of vegetal remains. The use of bean-like ideograms (thought by some archaeologists to have been suggested by the lima bean) has been reported from Lambayeque, as well as a species of bean (*Phaseolus* sp.) which possibly was used in certain games.

The Chicama, Moche and Virú Valleys constitute the first area from which satisfactory plant materials have been reported. The careful research of Bird, Larco Hoyle and Tello have furnished information on the ethnobotany of the cultures of the Chicama and Moche valleys, while that of Strong, Evans, Willey and Ford have supplied similar information for the Virú Valley to the south. Such data and the detailed descriptions of the remains of agricultural methods that were employed make possible a reconstruction of the use of wild and cultivated plants by the inhabitants in the course of their cultural history.

Epoch of Incipient Agriculture

The designations for the early agricultural pre-ceramic culture on the Peruvian coast appearing most frequently in the literature are "Incipient Agriculture" and "Early Farmers" (a term used by Bird, 1948a and b). Strong, in his chronology (Strong and Evans, 1952, Table 18), makes this an epoch which he calls "Incipient Agriculture," preceding the Formative. Willey and Phillips (1955) describe this manifestation as the "Preformative Stage." Some authorities mark the end of the Epoch of Incipient Agriculture and the beginning of the Formative by the first appearance of pottery. However, there are sites that appear to belong to a transitional stage and consequently are difficult to place chronologically. The culture represented is basically that of the Epoch of Incipient Agriculture, yet a plain, simple pottery is found either at the site or in the immediate locality. Donald Collier (1960 in ed.) has met this dilemma by subdividing the Epoch into a "Preceramic" and an "Initial Ceramic" stage. This treatment of the Epoch will be followed in the present work. Included in Collier's Preceramic designation are the periods of Huaca Prieta and Cerro Prieto; in his Initial Ceramic are the Early and Middle Guañape Periods.

Huaca Prieta Period. The site of Huaca Prieta is located at the mouth of the Chicama Valley near the small, modern settlement of El Brujo. It consists of the large midden of Huaca Prieta and a slight elevation of debris that rises to the north. The Huaca, as its name implies, consists of black earth, the result of the accumulation of debris over a long and continuous occupation. Until 1946, this site, though familiar to archaeologists, had received only cursory study. In that year Junius Bird conducted his painstaking excavations (Bird, 1948a, b; Bennett and Bird, 1949). The results were surprisingly rewarding. A large number of non-perishable and perishable specimens were recovered which represented three cultural periods, the pre-ceramic Huaca Prieta and the ceramic Early Guañape and Cupisnique. The former was found throughout the Huaca and in the lower levels of the slight debris accumulation to the north. At the latter location, the Early Guañape and Cupisnique cultures overlay the earlier levels, thus representing a continuous sequence. The absence of weapons at Huaca Prieta supports the belief that these fishermen-agriculturists were a non-hunting, peaceful people who occupied the site over a long period of time without apparent cultural change. Furthermore, the succeeding Cupisnique culture was the result of infiltration rather than force.

The earlier levels of Huaca Prieta show a small settlement of people living in semi-subterranean houses of one or two rooms made of cobblestones held together with a mud "mortar." Their material culture was indeed simple. Twined baskets, reed mats and some pounded bark-cloth have been found, as well as fish-nets and cotton cloth, the fibers of which were sometimes combined with another vegetal fiber. Percussion-flaked knives and scrapers, stone net-sinkers and an occasional bone needle are among the limited categories of tool specimens. The lack of pottery was probably met by the use of baskets and gourds for containers, both types having been found. These early agriculturists lived a precarious existence, depending for their subsistence chiefly upon seafood, which they supplemented by gathering wild food plants and practicing a primitive form of agriculture. The vegetal remains recovered from the site consisted of both wild and cultivated species in the form of actual plant parts and some artifacts of plant origin. The presence of this perishable material may be attributed to the fact that the site was located on a rocky plateau above the level of ground water. With the exception of the detailed study by Whitaker (Whitaker and Bird, 1949) of the gourd and other cucurbit remains and the recent discussion by Hutchinson (1959) of the cotton that was recovered, the lists of the plant species that have been published are only tentative. The current study of these vegetal remains by Towle (MSS) has thrown light on two other species that were present, namely, the lima bean (*Phaseolus lunatus*) and the chile pepper (*Capsicum* sp.)

That a simple type of agriculture was practiced by the early inhabitants of Huaca Prieta is evidenced by the occurrence of several species which in all probability were cultivated. These include the squashes, *Cucurbita ficifolia* and *C. moschata*, the common gourd, *Lagenaria siceraria*, and the lima bean. Four other genera, though known to occur in the wild state, may also have been tended or possibly planted. These are *Canavalia*, the chile pepper, cotton and achira, a plant which produces a starchy, slightly sweet rhizome. However, the most interesting feature of the discovery of this primitive agriculture is not so much the presence of these cultivated species as the absence of maize. Here was a pre-ceramic community practicing a form of agriculture before the advent of this important American staple food. In fact, maize does not occur at the site of Huaca Prieta until the Cupisnique period, when it appears in association with pottery and various new techniques of weaving. Among the remains of wild plants recovered from the earlier levels are the starchy rootstocks and tubers of the cat-tail and certain species of *Scirpus* and *Cyperus*. In addition, two trees (possibly cultivated) supplied edible fruits, the lucuma (*Lucuma sp.*) and *ciruela del fraile* (*Bunchosia*, probably *B. armeniaca*). The botanical identifications of several seeds, the tree which furnished bark for the pounded bark-cloth, and the plant fiber sometimes combined with cotton in the making of textiles, are yet to be determined.

Aside from the remains of plants that were cultivated to a greater or lesser extent during this Epoch of Incipient Agriculture, one must look further for evidence of agriculture. Bird does not mention in his tentative resumes of the work at Huaca Prieta the discovery of either crude hoes or digging sticks, nor are such specimens recorded at the contemporaneous site of Huaca Prieta de Guañape, in the Virú Valley. The early farmers may have used scrapers of stone or of gourds to loosen the top soil before planting or as cultivators, but such specimens, though showing usage, would not necessarily point directly to their having been agricultural tools. The same may be said of the simple digging sticks which had not yet been developed into a characteristic form of implement. At the time the site of Huaca Prieta de Chicama was occupied, the river probably formed lagoons and swampy areas along its course which supported a *monte* vegetation. Under such conditions a simple type of garden agriculture would have been possible.

Cerro Prieto Period. As mentioned above, the period representing the pre-ceramic stage of the Epoch of Incipient Agriculture in the Virú Valley is known as Cerro Prieto, a name derived from a small peninsula of black rock not far from the modern fishing settlement of Guañape. Three sites, located near the coast, are definitely assigned to this culture, while two others, farther inland, although still considered doubtful as to their cultural identity, may also belong to this early epoch (Willey, 1953, pp. 38–40). One of the sites in the lower valley is Huaca Prieta de Guañape (sometimes referred to as Huaca Negra; Strong and Evans, 1952, p. 7), which was excavated in 1946 by Strong and Evans and by Bird. A pre-ceramic culture affiliated to that found at Huaca Prieta de Chicama was discovered here. This cultural resemblance is based upon similarities of building structures, the occurrence of textile fragments of twined cotton and another plant fiber, and the cultivation of cotton and gourds. These two sites further resemble one another in that each demonstrates an uninterrupted cultural sequence which extends from the pre-ceramic early agricultural epoch, in which maize is lacking, into a later ceramic epoch in which maize occurs. This later period is designated as Cupisnique in the Chicama Valley and as Middle Guañape in the Virú.

The site of Huaca Prieta de Guañape is composed of three mounds, none of which is as large as that of Huaca Prieta de Chicama. These mounds are made of sand, and since the ground-water level fluctuates in that part of the valley, little in the way of perishable remains has been preserved. This site, like others in the Epoch, is one of a small settlement of people living in simple, subterranean houses made of packed clay with roof-beams of algarroba or possibly of canes. They depended largely upon the sea for their subsistence, as is shown by the presence of the bones of large fish and numerous shells of mussels, sea snails and rock crabs. However, this seafood diet was extended at least to some extent by the use of wild and probably cultivated food plants. Although the site is not conducive to the preservation of perishable remains, the presence of both cotton (*Gossypium barbadense*) and gourds (*Lagenaria siceraria*) indicates that some form of agriculture was practiced. One may refer to the list of plants represented in the early levels at Huaca Prieta de Chicama for other cultivated as well as wild species that were in all probability also used at the Virú Valley site. To the specimens of cotton and gourds found at Huaca Prieta de Guañape should be added the vegetal remains of plants used in house construction, a specimen of chewed plant fiber, and the plant fiber that was sometimes combined with cotton in the making of textiles. Although specimens of agricultural tools have not been reported from Huaca Prieta de Guañape, we suspect that the type of agriculture practiced resembled that of Huaca Prieta de Chicama. Similar conditions may have existed in the lower valleys of both rivers, where marginal lands made possible a garden type of agriculture without the necessity of irrigation.

Guañape Period. This period has been designated as Guañape, from the type site of Huaca Prieta de Guañape where evidence of this culture overlies that of the pre-ceramic Cerro Prieto (Strong and Evans, 1952, p. 206). Some eighteen locations in the Virú Valley, including small scattered settlements, community or ceremonial buildings and cemeteries, have been identified as either definitely representing this period or containing pottery attributable to it (Willey, 1953, p. 44).

In the Virú Valley, the Guañape period was of long duration, and based upon pottery types and other cultural variations, has been divided into an Early, Middle and Late phase. Early and Middle Guañape sites are to be found on the coast, whereas those of the Late phase, which have no plant remains (Collier, 1955, p. 133) are located inland on the marginal land of the valley. Both the cultures of Early and Middle Guañape are represented at the type site. A similar succession is also demonstrated at Huaca Prieta de Chicama (Bird, 1948b, p. 26), where Early Guañape follows the pre-ceramic levels of Huaca Prieta.

The culture of Early Guañape, which Collier assigns to his Initial Ceramic stage, is set apart by the presence of a plain pottery and a noticeable improvement in techniques such as weaving. Guañape pottery is simple, red or black ware which, although considered of local manufacture, was probably not an invention of this coastal area. The communities continued to be small and depended largely upon the sea for their subsistence, a diet which we may safely assume was at least in part augmented by the use of wild and cultivated plants. Because of the poor conditions for the preservation of perishable remains at Huaca Prieta de Guañape, only a few specimens of plant origin were recovered. These include cotton textiles, the presence of which points to the practice of some form of agriculture. This supposition is strengthened by the occurrence of cotton and gourds in the earlier pre-ceramic levels of Cerro Prieto. However, remains of a wider variety of species were recovered in Early Guañape levels at Huaca Prieta de Chicama. A study of these is yet to be published.

With the appearance of Middle Guañape pottery in the upper levels of Huaca Prieta de Guañape, there is marked cultural advancement. However, as was the case in previous strata, the plant remains are largely restricted to cotton textiles. If we assume that some form of agriculture was practiced in the Cerro Prieto and Early Guañape cultures at Huaca Prieta de Guañape, then it seems probable that the same species continued to be cultivated during the Middle Guañape phase.

There is no mention in the literature of evidences of agricultural techniques at any of the Guañape sites except the observation by Ford and Willey (1949, p. 23) that some of the early buildings that were excavated in the Virú appear to have been constructed so as to catch and conserve flood-waters from the Virú River. However, the Guañape settlements were located on marginal lands near which there was a water supply sufficient to support small-unit farming (Willey, 1953, pp. 400–401). As mentioned above, a similar situation existed at Huaca Prieta de Chicama and the type of agriculture practiced at the two sites very probably had much in common.

Formative Epoch

Cupisnique Period. The first conclusive evidence of the Chavín Horizon on the North Coast comes from strata representing the Cupisnique Period, the radiocarbon date of which has been placed at approximately 750 B.C. This culture was first recognized at the type site of Cupisnique in the Chicama Valley, and although it appears to be a development of this area, evidence of this culture (with some variations) has been found at other sites in valleys to the north. The Cupisnique culture follows that of Early Guañape and possibly Middle Guañape at Huaca Prieta de Chicama, and it seems that no distinct break occurred in the cultural sequence from the pre-ceramic Huaca Prieta Period to that of Cupisnique. This continuity suggests a peaceful infiltration of immigrants who brought with them a culture containing elements new to the area. The communities remained small and consisted of scattered houses made of stone and cylindrical adobes built above ground. New techniques of weaving were introduced but these did not supplant the older methods, thus pointing to a gradual fusion of the new and old cultures.

Like the people of the preceding periods, those of Cupisnique times depended upon both the sea and wild and cultivated plants for their subsistence, a diet that was increased in the latter part of this period by hunting. Plants that were found in the pre-ceramic levels at Huaca Prieta de Chicama are also represented in the Cupisnique strata. Specimens of *Lagenaria* are abundant, and both shells and seeds of a variety of warty squash (*C. moschata*) and the seeds of a variety of *C. ficifolia* were recovered (Whitaker and Bird, 1949, p. 14). The most important discovery, however, was that of maize in a level containing a slightly more advanced type of pottery than the Early Guañape ware, one that corresponds to the pottery recovered in the Virú Valley from levels of the Middle Guañape phase. This maize is a small-eared type with tunicate characteristics which probably indicate an early pop corn. Similar maize has been found at the Aspero site near Puerto de Supe to the south, in levels that belong to the Pre-Ceramic or early Initial Ceramic stage of the Epoch of Incipient Agriculture (Towle, 1954, p. 135; Willey and Corbett, 1954, p. 29). The list of plants was further increased by finding the remains of the peanut, of a species of *pacae* (of which there are probably two varieties), the avocado and a brown cotton (Bird, 1948b, p. 27). It is interesting that some evidence has been found which suggests the possibility of manioc having been grown at this time for its edible roots (Bennett and Bird, 1949, p. 126).

Pottery became defined and elaborated. An interesting as well as instructive development in the ceramic art was the modeling of human, animal and plant representations. The first naturalistic forms appear, and depictions of the plant parts that were used for food are common. These include the peanut, yuca, gourds and a fruit identified by Larco as a cultivated form of *caigua* (Larco, 1946, p. 150). According to Herrera (1939, p. 129) this name applies to *Cyclanthera pedata.*

With this increase in the importance of agriculture, one would expect to find evidence of agricultural techniques. Such information is lacking, however, as it was in the pre-ceramic and Guañape levels. The use of small garden plots, laid out in marginal lands where irrigation was not necessary, was in all probability the common agricultural pattern. Such lands were available along the river, not only to those people living in the lower Valley but also to those inhabiting sites farther inland.

Salinar and Puerto Moorin Periods. The subsequent Salinar Period in the Chicama Valley and the contemporaneous Puerto Moorin Period in the Virú add little to our knowledge of either the cultivated plants or the agricultural methods. Although we lack definite information on Salinar settlements, the villages were presumably located on marginal lands as were those of the previous Cupisnique Period. However, the pottery representations of house-types show a definite architectural development over those portrayed in the ceramics of the earlier period. The people further resembled those of the Cupisnique in that their food economy was largely dependent upon domesticated plants.

Maize, squash seeds, and gourds, some of the latter filled with material which may have been food, have been found in graves. Molded pottery forms were common in Salinar ceramic art and give further clues to their ethnobotany. Such representations as have been identified include the cactus and two edible fruits, the lucuma (*Lucuma* sp.) and the *pepino.* This latter vernacular name is used to designate the fruits of two different native plants, *Solanum muricatum* and *Cyphomandra splendens.* We may safely assume that other plants might be attributed to these people, since there is the possibility that additional genera may be added by finding ceramic representations or even actual plant remains. To this vegetable diet was added sea food. The presence of bones of birds and of the llama suggests that these animals also were possible sources of meat.

We have no evidence upon which to reconstruct their agricultural methods, but the picture implies an imperfect control of agricultural techniques, including irrigation, if it was practiced at that time (Willey, 1953, p. 400).

More satisfactory information is available on the Puerto Moorin Period than on that of Salinar due largely to the detailed settlement pattern studies made by Willey (1953). Eighty-three sites have been identified as belonging to this period, the greater number of which are concentrated on the Huacapongo River, a branch of the Upper Virú. Those sites in the Lower Valley have a scattered distribution similar to those of the Guañape Period. However, late in the Puerto Moorin, there was apparently a population shift from the marginal lands of the upper Valley to the lower coastal areas. Building structures of the period reflect an embryonic socio-political tendency that was to be demonstrated in the Gallinazo Period by pyramids and community dwellings impressive in number and size.

The size and complexity of the Puerto Moorin sites suggest an increase in population over that of the Guañape Period. This may well have been stimulated by the cultivation of maize and the development of agricultural methods. The first evidence of maize in the Virú Valley occurs in this period. However, no remains of canals or aqueducts have been found. It has been observed that this was a time in the prehistory of Peru of experimentation in new techniques and controls, which probably included the beginnings of organized agriculture.

This was to flower in the later Gallinazo Period as an intensive agriculture supported by large irrigation systems.

The evidence of plant remains from this period is limited. The cemetery excavated by Strong and Evans (1952, pp. 47–59) has thrown some light on the culture of the people of the Valley at this time. Among the grave finds in the simple pit burials are gourd dishes and a few gourds containing remains of maize and peanuts. Mats and cordage of twisted fibers and cotton cloth were used in the preparation of the body for burial. In one grave a staff of algarroba wood had been placed. An interesting specimen recovered was a flanged, stone mace-head, the shape of which was reminiscent of the fluted cactus common in coastal Peru. A small, old break had been repaired with resin, a method of mending that was sometimes employed in pottery recovered from later levels. This specimen was not included in a grave and may possibly have become lost even before the Puerto Moorin Period.

Although we have no knowledge of the agricultural techniques that were employed at this time, the remains of cotton, peanuts, gourds and maize point directly to the practice of a form of agriculture that possibly had already become established as part of the general cultural pattern. In addition to the domesticated plants, the occurrence of remains of wild species shows that the botanical environment of the Virú Valley still supplied raw materials for industrial purposes.

A number of well-known sites attributable to the Formative Epoch and possibly contemporaneous with the Salinar and Puerto Moorin Periods are located in the area of the Santa (Chimbote), Nepeña and Casma Valleys. These are noteworthy because of the temples, pyramids, and other large building constructions, some of which are embellished with designs and sculptures in relief. The importance of these sites lies in their Chavín affiliations which are apparent in both construction methods and art forms. Although these locations have received considerable attention from archaeologists, there is a paucity of information on the vegetal remains that may have been in association with other materials.

Gallinazo Period. The cultural trends of the Puerto Moorin Period in the Virú Valley continued and developed during the Gallinazo, a period of considerable length that extended from the Late Formative well into the Classic Epoch. The period has been divided into an Early, Middle and Late phase, based upon ceramic types and percentages; the entire period, however, is identified with the Negative Painting Horizon, a style native to the Virú Valley. The same types of sites that characterized Puerto Moorin are to be found, with, however, a marked increase in their number, size and complexity. This period was one of great construction activity. It is of ethnobotanical interest that the adobes used during the period were apparently made in cane-lined moulds. The development in building and architecture reflects an increase in the population of the valley, which is thought to have reached its maximum during Gallinazo times.

The subsistence economy of this large population was based upon intensive and organized agriculture, supported by a planned irrigation system. Canals, the remains of which are still traceable, led from the upper parts of the river to the lower valley. The large and impressive structures known as "castillos" are thought to have been built for the protection of the irrigation system. It is possible that the canals may have been begun in the Puerto Moorin Period, which would conceivably have contributed toward the increase in the population in the Huacapongo basin during that time (Willey, 1953, p. 419). All arable land was watered and cultivated (*ibid.*) and irrigation plots are still evident to the south and southwest of the type site of Huaca Gallinazo (*ibid.*, p. 133).

The common domesticated species that are represented among the plant remains of the Puerto Moorin Period are also recorded from Gallinazo levels in the Virú Valley. These include cotton, maize, peanuts and gourds. In addition, remains of the crookneck squash (*Cucurbita moschata*) were found at Castillo de Tomaval, as were those of the edible fruits of the lucuma, avocado and guava. The Gallinazo crafts drew heavily upon the natural botanical environment. Both worked and unworked specimens of wood are commonly encountered. Strong and Evans (1952) mention finding wooden stoppers, cordage, baskets, matting and an algarroba digging-stick among the artifacts of plant origin. The latter is of particular interest because of its agricultural significance.

A corn-popper and a molded stirrup jar (*ibid.*, pl. 7, figs. g and h) accompanied a burial at Huaca Gallinazo. The latter vessel is supported by outwardly curved feet and resembles others of the type found among the ceramics of the Chicama Valley. The supports are thought to have been modeled after curved tubers.

Our knowledge of the occurrence of Gallinazo culture in the Chicama region is still incomplete. Salinar was followed by a poorly defined period characterized by the presence of Negative Painting pottery, a type particularly associated with the Gallinazo Period in the Virú Valley, and unquestionably Gallinazo burials have also been found. Whatever its status, the Gallinazo Period in the Chicama Valley preceded and contributed toward the formulation of the Mochica culture of the Classic Epoch.

Classic Epoch

Mochica Period. This period constitutes the Classic Epoch in the Chicama and Moche Valleys and represents the peak of mastery of the various techniques that had been developing during the Formative. Mochica stands as a counterpart of the Nazca culture of the South Coast, and like it, produced a representative art style that sets it apart from other regional manifestations.

The Mochica art style is characterized not only by the two-dimensional painted designs of its bicolored pottery but also by its sculptured and molded ceramics. These latter forms demonstrate a realism that has not been surpassed in New World art. The designs of both the painted pottery and those in the round often depict the daily life of the people, this genre art furnishing us with a pictorial account of many phases of Mochica culture including their economic plants.

Although the center of development of Mochica was in the Chicama Valley, its influence extended to the Moche, Virú, Chao, Santa (Chimbote) and Nepeña Valleys. The Huancaco Period of the Virú is the best known of the peripheral manifestations of Mochica culture. Larco (1946, p. 161) recognizes four stages in the Mochica Period, the earliest two being confined to the Chicama and Moche Valleys, the latter two extending their influence over the maximum Mochica territory.

There is no recognized type site for the period. In fact, there are no remains of large villages and only a few of small communities. It appears that the inhabitants lived in settlements near the cultivated areas. However, little is known of the type of dwellings that were used.

In contrast to this limited information regarding the habitations of the Mochica, we have the remains of massive structures of ceremonial centers, pyramids, aqueducts, roads and other public works. There is a similarity between these and the large constructions of the Gallinazo Period in the Virú Valley (Willey, 1953, p. 409).

Although the subsistence economy of the Mochica Period was based mainly upon agriculture, fishing, hunting and the gathering of wild food plants also contributed toward the food supply. The Mochica people not only cultivated the fertile valleys but the drier marginal lands as well, the latter receiving the necessary water supply from an extensive irrigation system. On the valley floor, the level areas were divided into small plots or fields outlined by banks of earth and marked in straight furrows through which the irrigation water flowed (Larco, 1946, pp. 162–63). On the sloping ground back from the river valley, similar ditches followed the contours of the terrain. In the valley, the flooding of the fields over a period of years leached the mineral salts contained in the soil and those formed a hard crust. This layer was in all probability removed by hand, since the digging stick might well have proved too fragile to break and lift the hard earth. The discarded surface soil was apparently piled to one side in the field; similar piles or small mounds have been observed in the Virú Valley. The hoe and digging stick were the principal tools used in cultivating the crops. Specimens of both implements have been recovered, the latter usually fashioned from a branch of algarroba. Bird guano, large deposits of which are found along the Peruvian coast, was used to fertilize the fields (Larco, 1946, pp. 162–63).

The irrigation system consists of canals and aqueducts and shows evidence of careful planning. Probably the most arresting of these irrigation projects in the Chicama Valley is the *Ascope* aqueduct built across a wide-mouth quebrada for a distance of some 1,400 meters. The embankment, along the top of which runs a canal, is constructed of earth and adobe to a height of fifteen meters. Still another impressive construction is *La Cumbre,* a canal which carries water from the headwaters of the river to the irrigation ditches of the fields about Chanchan, 130 kilometers distant. Other aqueducts and dams attributable to the Mochica Period are to be found in the Moche, Santa and Nepeña Valleys. Although the irrigation system of the Chicama Valley was in use during the Mochica Period, the time of its construction remains obscure, due in part to our limited knowledge of the earlier Gallinazo Period in this area.

The food plants cultivated within the area of Mochica influence consisted of:

maize	llacon
common beans	achira
lima beans	crookneck squash
peanuts	Hubbard squash
sweet potato	pepino
white potato	gourds, possibly used for food
manioc	as well as in industry
peppers	

A number of fruits were also used; some of these may have been grown or at least given some horticultural care:

chirimoya	guava
guanabana	lucuma
pacae	ciruela del fraile
tumbo	papaya
granadilla	pineapple
Prosopis sp.	

The masticatory, coca, was widely used, as is attested by painted scenes of coca chewers on jars as well as by molded vessels depicting men with distended cheeks and a small woven bag for the leaves at their sides. Although this plant grows wild in Peru, it is probable that it was also cultivated. There are representations of medicine-men or physicians treating patients, suggesting the possibility that the Mochicans used certain herbs which they thought possessed curative properties. Larco (1946, p. 163) mentions three such plants that are employed in folk-medicine today and were probably used then, namely *habillas, ashango* and *maicheles.* The first of these vernacular names is given by Herrera (1939, p. 189) as applying to *Hura* sp. or to *Sicydium diffusum.* There are still other wild species depicted on pottery, among them certain cacti, *caña brava,* and characteristic plants of the landscape. Many of these were probably used for specific purposes as well as artistic motifs.

From the available data, it appears that the Mochicans, like the people of preceding cultures, also depended upon the local flora for the raw materials needed in manufacturing. Cotton and gourds continued to be the cultivated industrial plants, while stands of *Scirpus* or other fiber-producing species may have received some care in their native environment.

The list of Mochica economic plants is derived largely from painted ceramic designs or from molded or modelled pottery forms. Edible plant parts, such as fruits and tubers, were reproduced as single vessels made from a mold of the original specimens, whereas plant parts too small to be of practical value when treated alone become units in a plastic composition. Thus jars in the form of squashes, tubers and ears of maize are common, while beans, peanuts, peppers and small fruits are usually combined to form larger vessels. The same realism that characterized other types of Mochica ceramics is found in the plant reproductions. The regard for detail on the part of the Mochica craftsman facilitates, in the majority of cases, the identification of these plant forms. Published accounts of the vegetal remains recovered from archaeological sites, however, will add to our present knowledge of Mochica ethnobotany.

From the agricultural evidence and the large amount of Mochica archaeological material, it might be assumed that the population of the Chicama and Moche Valleys was reasonably large. However, there is little evidence of any highly organized society (Kroeber, 1948, p. 830), and the degree of their social and political organization is difficult to ascertain. Nevertheless, the large constructions and ambitious irrigation systems do point to a systematic organization of manpower. How such communal enterprises were conducted remains debatable. That Mochica society had a ruler, apparently of divine origin, has been suggested (Larco, 1946, p. 167). This personage often appears in pottery representations and graphic art, where he is shown as an anthropomorphic, feline deity characterized by great fangs. He appears as a hunter, as a fisherman, and as an agriculturist presiding over agricultural ceremonies. The effigy on the digging stick found in the Warrior-Priest's grave of the Huancaco Period in the Virú Valley must certainly be his representation. It is of special interest that zoophytomorphic figures, which possess some of the more prominent of his characteristics, are depicted in association with domesticated food plants. All in all, the importance of agriculture in the thought and emotions behind Mochica culture is demonstrated in both their painting and molded ceramic art. Not only are domesticated and wild food plants frequently represented but the depiction of ceremonies connected with agricultural practices in association with the fanged deity is strong evidence of the respect and reverence in which their agricultural activities were held.

Huancaco Period. The Gallinazo Period in the Virú Valley was apparently terminated at the time of conquest by the Mochican people (Strong and Evans, 1952, p. 216; Willey, 1953, p. 31). The succeeding cultural period of Huancaco derives its name from the type site located in the southwest section of the valley.

There was a decrease in building activity. The characteristic cane-marked adobes of Gallinazo construction were replaced by small, rectangular ones without cane impressions. From the study of the distribution of Huancaco sites in the Virú, it is apparent that there was also a change in population density (Willey, 1953, p. 419). All of the valley was now inhabited. The subsistence economy of the people, however, was the same in both the Gallinazo and the Huancaco periods. The intensive agriculture which became well established in the Gallinazo continued during the Huancaco and there are no marked differences between the two. It is probable that the same pattern of irrigation also extended from one period to the other. There is evidence that the north and the south canals were in use during both periods (Willey, 1953, p. 363). The north canal and its distributary ditches supplied water to the extensive cultivated areas north of the river, the smaller branches of the network often returning to the main canal when gravity allowed. The south canal on the opposite side of the valley furnished water to the adjacent communities, some of which were near the sea at some distance from the river. The remains of "pukios" have been observed at sites behind the beaches (Strong and Evans, 1952, p. 9). These plots are shallow rectangular excavations that were made in order to reach the ground water level. Willey (1953, p. 364) mentions a site near the beach which has remains of "clay-crackled" cultivation plots some 20 meters square that were fed by a canal. Small mounds scattered over the region probably were formed by heaping up the salt-encrusted or salinated soil during prehistoric times.

The wide use of plants in the material culture of the Huancacoan people is demonstrated in the number and variety of such specimens recovered from burials. In addition to the more common artifacts of plant origin, such as mats, basketry, textiles, gourds, etc., digging-sticks and ceremonial paraphernalia are found. The unique tomb of the Warrior-Priest discovered at Huaca de la Cruz is of outstanding importance. This burial was that of an old man of great prestige, who must have symbolized the earthly representation of the tusked warrior-god, *Wira-Kocha* or *Ai apaec*. The study of this burial with its rich assortment of grave accompaniments offers many suggestions of the part played by a person of high position in the agricultural life of the people (Strong and Evans, 1952, p. 196). Among the symbols of rank is a ceremonial digging-stick surmounted by an effigy of the tusked god, accompanied by a small boy who holds three pieces of turquoise in his left hand; these are probably symbolic representations of maize kernels (*ibid.*, p. 157). Strong (*ibid.*, p. 199) calls attention to the similarity between the part played by the Warrior-Priest in starting the planting season and the later Inca ceremony in which the Inca or his representative initiated the planting season by breaking ground with his golden plow.

There is little doubt that the Mochica-Huancaco culture of the Chicama-Moche Valleys and the outlying provinces, such as the Virú, represents the peak of agricultural development. Few species were to be added to the list of cultivated plants in subsequent periods. Judging from the remains of domesticated species found in the Huancaco levels in the Virú, as compared to those found in Gallinazo, no new genera are represented. The list includes cotton, maize, gourds, squash (*C. moschata*), common beans, and lima beans. Species that may have been cultivated to some degree were *Canavalia,* lucuma, avocado and the guava. An effigy jar of a human head made in the shape of a tuber of the potato (*Solanum* sp.) was discovered in the tomb of the Warrior-Priest (Strong and Evans, 1952, pl. 28, fig. c). The "eyes" of the potato are clearly molded, as in similar jars from Mochica sites in the Chicama-Moche region.

This list of cultivated and wild species of food plants is small compared to those identified with the Mochica culture of the Chicama-Moche Valleys. However, the close connection existing between the Mochica and Huancaco cultures leads one to suspect that the Virú list is smaller than it need be. Since the source material for the present study rests mainly upon actual plant remains and plant representations, we can only hope that subsequent excavations will bring to light still other species which were used during the Huancaco Period.

Pottery in plant forms of Mochica origin have also been found in the upper levels of sites in the area south of the Virú Valley. These specimens include representations of pacae, achira, yuca, pepino (*Solanum muricatum*), potato and possibly *Caryocar amygdaliferum,* all of which were found in the vicinity of Chimbote (Santa).

Fusion Epoch

The Classic Epoch is brought to a close with the appearance of the strong influences of the Tiahuanaco Horizon. The succeeding Fusion Epoch represents an intermingling of Mochica and Coastal Tiahuanaco traits. In the Virú, this Epoch is described as the Tomaval Period. It was an era of apparent unrest throughout most of the Central Andean region and this was reflected in both its political and its social aspects. Elaborate planned communities consisting of large enclosed compounds with many rooms, referred to as "lay-centers" (Schadel, 1951), made their appearance. Such communal buildings are well demonstrated in the Virú Valley.

Intensive agriculture continued to be practiced. The upper Virú Valley, and extensive areas near the coast as well, were cultivated during both the Tomaval and the subsequent La Plata Period; the distribution of water by canals and irrigation was similar to that in the earlier Huancaco times. Willey (1953, p. 265) calls attention to the remains of walls that were built in the Huacapongo Valley to protect the immediate area and the whole valley from attack by people from the inland mountain gorges. Such precautions were deemed necessary to safeguard the canal system upon which the agriculturists of the valley depended.

Epoch of Kingdoms and Confederacies

The Tiahuanaco Horizon was not of long duration, the succeeding periods of the Epoch of Kingdoms and Confederacies being marked by the re-emergence of local styles. Two periods represent the Epoch in the Chicama Valley sequence: the Black-White-Red, which derives its name from the dominant ceramic style, and the Chimu. True cities appear on the North Coast at this time, the great city of Chanchan in the Chicama Valley having been built during the Chimu, which witnessed the growth and dominance of the Chimu kingdom. Such concentrations of the population suggest a well-organized social and political society, a fact that is confirmed by the reports of the Inca and Spanish chroniclers. The ruins of these cities include remains of irrigated areas and reservoirs, and these, with the basic irrigation system inherited from the earlier Mochica Period, formed a well-knit agricultural pattern. The ethnobotany of the people of the Chimu culture as well as that of the preceding Fusion Epoch was similar to that of the Mochica, with the possible addition of certain cultivated and wild species common to the highland region.

In the Virú Valley, there is no apparent cultural break between the Tomaval and the La Plata Period, at which time the Valley came under the domination of the Chimu kingdom. The general agricultural practices appear to be the same and the defense walls of the Huacapongo Valley continue to be used. This defense area reminds one of the site of Galindo above Chanchan, whose population defended the lower Chicama Valley with its great city. In fact, the site of Chanchan is similar in many respects to the smaller Virú Valley sites (Willey, 1953, p. 416). Both the city of the Chicama and the villages of the Virú are located near the sea. The pattern of cultivated plots situated near the dwellings is similar in both; even the S-shaped garden areas and *pukios* or *hoyas* were used.

Imperial Epoch

After the Inca conquest and the onset of the Imperial Epoch, or the Estero period of Virú chronology, the general cultural picture remained much the same, although archaeological materials of Inca origin and influence are to be found throughout. The fact that there was not more fusion between the two cultures may be due to the short period of Inca domination, which was terminated by the coming of the Spanish.

It is likely that the agricultural pattern and the canal system of the preceding periods persisted. Although there are few plant remains from the Tomaval, La Plata and Estero Periods of the Virú, it is safe to assume that those domesticated plants and wild species that had become an established part of the ethnobotany of the people of the Gallinazo-Huancaco Periods continued to be widely used. However, with the advent of both the earlier Tiahuanaco and the Inca cultures, there is a strong possibility that species of plants native to the highlands may have found their way into the Virú Valley culture. Such evidence may yet appear as vegetal remains or even artistic representations.

THE CENTRAL COAST

This geographic division of the Peruvian coast is separated from the North Coast by the Huarmey River and from the South Coast by the Lurín. In this area approximately eight rivers, rising in the Andean foothills, flow through narrow valleys to empty their waters into the Pacific. The terrain of the region is similar to that of the North Coast, where the edges of the fertile valleys lose themselves in the intervening desert. Unlike the North Coast, however, the Central Coast (with the exception of the Huarmey Valley) possesses what appears to be a definite archaeological unity. The apparent schism between the Huarmey, the northernmost valley, and those to the south may be explained at least in part by the geographical isolation of the former. In fact, it has been suggested that this valley may prove to constitute a separate subdivision of the Central Coast (Bennett and Bird, 1949, p. 98).

The Central Coast is rich in archaeological sites, ranging from the remains of small, primitive communities, which relied mainly upon the sea for their subsistence, to great constructions of adobe and stone which indicate an organized society with a well-established agricultural economy. Many of the sites have been the subject of exploration and study, the greater part of our information being derived from the published reports of research done in the valleys of the Chancay, Supe, Ancón, Chillón, Rimac and Lurín. Although Reiss and Stübel conducted explorations at the Necropolis of Ancón during the latter part of the nineteenth century, it was the great German archaeologist, Max Uhle (1913, 1925), who, during his researches on the Central Coast in 1903–1904, first sensed the presence of cultural continuity. He observed that a white-on-red pottery was found in levels beneath strata characterized by the presence of a distinctive ware decorated with designs of interlocking motifs of fish and feline heads. This observation gave the archaeology of the Central Coast the first tangible basis upon which a cultural chronology could be constructed. Subsequent research has led to the formulation of such a chronology, which extends from the Epoch of Incipient Agriculture to the complex culture of the pre-Inca and Inca Periods.

Our knowledge of the use of wild and cultivated plants by the peoples of the region is dependent largely on the plant remains discovered from sites in the valleys mentioned above. Some of the reports of these explorations present the data in detail, while others give only slight mention of plant material. However, such information as we have at our disposal is presented below.

Epoch of Incipient Agriculture

The Epoch of Incipient Agriculture has been recognized at several locations on the Central Coast.

The Aspero Midden in the Supe Valley described by Willey and Corbett

(1954) has been placed in the Initial Ceramic stage of this Epoch. This site consists principally of a large midden with the remains of a structure near the edge of the area, and a cemetery. The walls of the building which are still traceable on the midden itself are often designated in the literature as the "Aspero Temple," because of the pounded clay or burned floor or platform which might have been used for sacrificial fires. It was beneath this floor that a cache of maize containing forty-nine whole and broken cobs was discovered. These specimens represent a type of corn similar to that found in the Cupisnique levels at Huaca Prieta in the Chicama Valley (Towle, 1954). No pottery was found at either the midden or in the "Temple" structure but the presence of red and black pottery in the Aspero cemetery as well as in two small adjacent middens seems to place the site in this latter part of the Epoch of Incipient Agriculture.

The excavations at the Aspero cemetery yielded baskets, mats of cat-tail and sedge, cotton textiles and gourd containers. The latter were in some instances filled with raw cotton or bivalve shells.

Evidence of a canal which suggests the use of irrigation at the site are still to be seen at Aspero. This canal was possibly part of a local system which included a second canal that furnished water to the other settlements in the area. However, the dating of these irrigation works is yet to be determined. Stumer (1954c) mentions the existence of sites in the Rimac Valley belonging to the Epoch of Incipient Agriculture. However, he describes only one, "Pedreros B." Although plant remains are not mentioned, the forthcoming detailed report of the excavation will doubtless include whatever vegetal material was found.

Still other locations that may throw light on the Epoch are the low shell mounds and refuse accumulations on the coast between the Lurín and Chilca Rivers. The presence of the simple pottery characteristic of the Early Ancón Period leads to the expectation that evidence of a still earlier ceramic agricultural period may be found. Although the culture of these earlier inhabitants of the Central Coast remains to be formulated and described, the available evidence suggests that, if not identical, it is in all probability allied to that of the North Coast.

Formative Epoch

Early Ancón Period. This culture, discovered by Uhle at the shell heaps at Ancón and the sites near Puerto de Supe, was recognized by him as earlier than those cultures previously described by Reiss and Stübel (1880–87) from Ancón. This early manifestation of a fishermen-farmer people is now designated as the Early Ancón culture, the earliest period of the Formative Epoch and the first to be well-defined in Central Coast chronology. Early Ancón is estimated to have flourished during the latter part of the first millennium B.C. (Willey and Corbett, 1954) and was contemporaneous with the Cupisnique Period of the North Coast. These cultures are similar in many respects, and all possess specific

Chavín characteristics. A number of sites attributable to the period are known to exist. Those located at Puerto de Supe, Ancón and Curayaco have received careful study and our best picture of the culture is derived from them.

Of the several sites in the Supe Valley which are assigned to this period, the Lighthouse site and a midden, Li–31, have been excavated. These sites are situated near the modern port of Puerto de Supe and near the Initial Ceramic site of Aspero.

The Lighthouse site (Willey and Corbett, 1954), consisting of an occupational area and a cemetery, is located in a narrow valley opening upon the sea and just west of Puerto de Supe. Pottery is plentiful at both units of the site, and there is little question but that they are affiliated. The dwelling area is limited in extent and is characterized by remains of shell middens and stone enclosures. Plant remains were recovered from burials in the cemetery. A wooden spatula and the teeth of two combs are associated with the skeletons in three of the graves. It apparently was customary to wrap the flexed bodies in cotton textiles and place reed mats beneath or above the body. Among the specimens found in the burials are pottery vessels, baskets, cordage and gourd containers, many of which contain burial offerings, such as raw cotton, maize and peanuts. It is interesting to note Uhle's reference (1925, p. 262) to finding lima beans, manioc and certain unidentified seeds at either the Lighthouse site or possibly at Aspero. Although no wild food plants appear to have been found, such native species were in all probability used, as they were at Huaca Prieta de Chicama.

The shallow midden, Li–31, although adjacent to the Aspero site, is apparently more closely related culturally to that of the Lighthouse and thus has been assigned to the Early Ancón Period. Specimens of plant origin are limited to a fragment of a cotton string fishnet. An interesting feature of this site is the prehistoric irrigation canal that crosses the area in which it is located.

In the small valley of Ancón, south of Supe and just north of the Chillón Valley, there are two sites: the Ancón shell heaps (Willey and Corbett, 1954) of the Early Ancón Period and the Necropolis of Ancón which belongs to later periods. These sites are located near the modern coastal town of Ancón in an area that has supported fishing communities since prehistoric times. The shell heaps point to the existence of an old fishing village that depended largely upon marine food for its subsistence. Rocks that may possibly be the remains of houses are scattered over the surface of the mounds, and a few burials have been found within the shell and ash refuse. A cemetery of the Early Ancón Period that apparently was associated with the settlement has recently been discovered nearby (Carrion, 1948). No remains of food plants have been reported from either the village site or the cemetery, whereas sherds, stone and bone artifacts and baskets have been recovered. The absence of maize and other commonly domesticated plants is puzzling, since specimens of these have been found in contemporaneous strata in the Chicama, Virú and Supe valleys. It is probable that the inhabitants of the Ancón village used at least some cultivated plants to augment their seafood diet. Possibly excavations at the cemetery will supply the answer to this question.

The economy practiced by the inhabitants of these Early Ancón sites in the Supe and Ancón Valleys depended upon the sea and upon agriculture. The Lighthouse site, Li–31 and the Ancón shell heaps are all in close proximity to the ocean, where marine food is plentiful. The presence of sea shells in the middens and specimens of fishing equipment furnish evidence of their exploitation of this source of food. Since remains of cultivated plants have been found at Aspero and the Lighthouse site, one also looks for evidence of the practice of agriculture. As Willey has pointed out (1954), these locations were either immediately adjacent to what is now arable land or not more than two kilometers from it. It was possible for the inhabitants of Aspero to have farmed the well-watered lands of the valley of the Supe that were located at the mouth of the quebrada in which the site is located, and although the Lighthouse site is in the desert, the fertile valley of the Pativilca River where farming could have been conducted without much difficulty is only about one-half kilometer away. However, the remains of two canals at the Supe sites suggest the possibility of irrigation. At the Aspero site, such a canal is still traceable, and it is apparent that the slopes overlooking the quebrada were cultivated at one time. Furthermore, at the midden site of Li–31 another canal can be traced as it follows along the hills, and it might have been part of an irrigation system. There is little doubt that these canals are prehistoric, but whether they were used in the Early Ancón Period or later remains questionable. From the information that has been obtained in the Virú Valley, it is thought that irrigation by canal probably did not begin until the Middle Formative.

If the present arid conditions in the Ancón Valley are representative of prehistoric times, this area was not favorable to the practice of agriculture. There is little available water and this is limited to a spring and a few brackish wells. The water that could have been brought in would not have been sufficient for farming purposes, and no conclusive evidence of irrigation of any kind has been found. The nearest arable land is at present some ten kilometers distant, in the Chillón Valley. However, such land may have been nearer in prehistoric times. Perhaps either the climatic and physiographic conditions were different during the early occupation of Ancón, or the village depended upon the crops grown in the Chillón Valley for the vegetable part of its diet.

A large, Chavinoid site on the rocky peninsula of Curayacu fifty-two kilometers south of Lima has been recently explored by Engel (1956). The inhabited area of the site is located on the top of a hill and on the steep side that slopes to the sea. The upper levels of the deep refuse belong to the late periods of Central Coast chronology; beneath are abundant Chavinoid remains consisting of bones, stones, ashes, shells and seafood. Maize and an assortment of vegetal remains represent the plants used.

The simple life that is represented in the lower levels is reminiscent of the culture of similar sites to the north. This was a community whose subsistence was derived from the sea and from agriculture. The identity of the plant remains will not only add to our present knowledge of the distribution of those used during this early period but in all probability will add to the list of both cultivated and wild species. In fact, the results of the research at Curayacu will contribute greatly toward defining more clearly the culture of the Early Ancón Period.

Although the area of habitation lacks water at the present time, crops are grown in a canyon to the north of the peninsula. Engel suggests that the available supply of fresh water in prehistoric times may have since disappeared and that crops may have been cultivated at a distance from the village, as they are today. We know little of the agricultural methods employed by the people of the early periods of the Formative Epoch on either the North or Central Coast.

Baños de Boza Period. This period is characterized by the distinctive white-on-red pottery, the pottery style of the White-on-Red Horizon. The two type sites are Baños de Boza and Cerro de Trinidad, both of which are located in the Chancay Valley. Excavations by Willey (1943) at the former site showed that it had been occupied only for the duration of the Period. The one burial that was uncovered contained no plant remains except woven cotton. The site of Cerro de Trinidad, also excavated by Willey (ibid.), is situated nine kilometers west of Baños de Boza. The lower levels contained white-on-red pottery while the upper ones contained pottery of the subsequent Playa Grande Period. It appears that the site had been inhabited continuously during these two periods. The plant material that was recovered consists of a few wooden objects and fragments of woven cloth included among the scanty grave finds.

With the exception of cotton, we lack evidence of any domesticated plants that might have been used. However, we are safe in assuming that agriculture was practiced and that those staple crops which are known to have been cultivated during the preceding Early Ancón Period were also grown during this later period. Likewise, we have no clues as to the possible agricultural methods that were employed, but since this Middle Formative Period of the Central Coast was contemporaneous with those of Salinar and Puerto Moorin on the North Coast, we might reasonably expect to find a similar pattern of agriculture occurring during the Baños de Boza Period.

Playa Grande Period. This culture flourished in the Chancay, Ancón, Rimac and Lurín Valleys, and lasted for a long time. The period is characterized by a pottery decorated with a design style of interlocking fish and feline motifs, either alone or in combination with dots, circles, steps or similar line combinations. Although the Playa Grande culture was confined to the Central Coast, it possessed many similarities to the Gallinazo of the North. They were both local developments of long duration. The increased size of the living areas and the massive buildings that are present point to a marked increase in population over that of the immediately preceding periods.

The similarities between these two contemporaneous cultures in all probability included a parallel development of agriculture and agricultural methods. Although mention is made of the use of irrigation at Maranga (Jijon y Caamaño, 1949) during the Playa Grande, we lack substantial evidence for other large-scale agricultural practices similar to those in the valleys of the North Coast during Salinar and Gallinazo times. However, remains of several of the more common cultivated species that we have come to associate with an established agriculture have been recovered from Playa Grande levels on the Central Coast. Specimens of gourds, cotton, lucuma, and fragments of wood and caña brava were recovered in the refuse heap below the southwest terrace of the Temple of the Sun at Pachacamac (Strong and Corbett, 1943; Towle, 1948). Among the remains from the type site of Maranga, Jijon y Caamaño (1949) mentions both the common and the lima bean and maize, and both wild and cultivated species were represented among the vegetal remains collected by Stumer at the site of Playa Grande, also attributable to this period. These include maize, peanuts, cotton, gourds, pacae, lucuma and *Bunchosia* sp. (probably *B. armeniaca*). In addition, small branches of *Buddleia americana* were among the plant remains found at a burial of the Classic Maranga Period at Vista Alegre (Towle, 1958).

Classic Epoch

Maranga and Nievería Periods. No true cultural florescence marked the Classic Epoch on the Central Coast, as was the case in the areas to the north and south (Strong and Evans, 1952, p. 245). The Maranga and Nievería Periods, unlike the preceding Baños de Boza and Playa Grande, are not characterized by a distinctive ceramic style. Rather, they contained elements of the earlier Playa Grande pottery as well as some associated with Mochica and the later coastal Tiahuanacoid.

Little information on the ethnobotany of the Classic Epoch is as yet available. However, its apparent relationship to other coastal cultures of the Classic and Fusion Epochs leads us to assume that the same advanced agricultural pattern was present in the valleys of the Central Coasts as in other contemporaneous sites.

Subsequent Epochs

The cultures of the later Epochs are well represented at the famous sites of Pachacamac and the Necropolis of Ancón. These sites have furnished us with well-documented archaeological remains from which the cultural history of each has been determined. The wealth of material recovered has, in both instances, included a wide range of plant remains, which have been carefully studied. As mentioned previously, no new domesticated plants were cultivated on the coast after the Classic Epoch. A highland-coastal cultural interchange, however, is indicated by the presence of certain highland cultigens.

The Necropolis of Ancón is one of the two units of the Ancón site. The other, the Ancón Shell Heaps, belonging to the Early Ancón Period, was described above. The Necropolis is the better known of the two, the cemetery and the occupational area having been the source of both small and large collections since the middle of the last century. The first organized exploration of the site was conducted by Reiss and Stübel in 1880–87 (1880–87). Their large collection of archaeological remains, including plant material, was described in a series of well-illustrated folio volumes. In 1912, Uhle excavated in the Necropolis and occupational area, and it was from his ceramic collection that Strong (1925) formulated a cultural sequence for this section of Ancón. It was shown that Ancón, like Pachacamac, was a coastal Tiahuanacoid site; moreover, the stratification of the two is similar. However, it appears that Ancón was abandoned after the Inca conquest; just when this occurred is not known, but the date is certainly before the arrival of the Spanish. Pachacamac, on the other hand, was in existence after the conquest.

Following Strong's ceramic sequence, the earliest level found at Ancón is the Early Ancón Period. This is represented at the Ancón Shell Heaps and at one of the locations in the occupational area excavated by Uhle. This early manifestation is followed by the coastal Tiahuanacoid periods of Middle and Late Ancón, both of which are divided into an Early and a Late phase. However, at the Necropolis, Middle Ancón I marks the earliest culture that has been found. The recent excavations at the Necropolis by Willey (1943) have corroborated Strong's sequence, as well as adding important refinements to the cultural history of the site.

The first plant remains from a Peruvian coastal site to be identified were probably specimens recovered from the Necropolis at Ancón. During the last half of the nineteenth century several archaeological collections, in addition to the large one made by Reiss and Stübel, were taken to Europe. Both specimens of plant remains and artifacts of plant origin were included. One of the first of such collections to be studied was described by Saffray (1876). This detailed description of a Peruvian mummy bundle contains a number of identifications of plant material included in the burial, and although the author does not mention the site from which it came, it might have been from the Necropolis.

Three years later, Rochebrune (1879), a French botanist, identified plant remains included in the collections made by De Cessac and L. Sativier which were exhibited in Paris in 1878. These two collections contained a large assemblage of vegetal remains. Rochebrune describes fifty species under the headings, *Food and Medicinal Plants* and *Industrial Plants and Dyes*. It is unfortunate that the author in many instances fails to mention whether the specimen under consideration is an artifact of plant origin or the remains of the plant itself.

Rochebrune's study was followed by that of Wittmack (1880–87). The material used by the latter was recovered at the Necropolis by Reiss and Stübel, and his study was included in the formal report of the excavation. The results of both Wittmack's researches and those of Rochebrune were incorporated by the former in his *Die Nutzpflanzen der alten Peruaner* (1888), together with the identifications of plant specimens from the collections made by Commander Acland and Lieutenant Holland at Ancón.

Still other identifications of plant remains from Ancón were made by Costantin and Bois (1910) and later by Harms (1922) in his reappraisal of the previous studies of ethnobotanical remains from Peru. Some few additional specimens are mentioned by Safford (1917a) and by Yacovleff and Herrera (1934–35). These studies combine to furnish us with one of the most complete descriptions of the plant materials from any Peruvian archaeological site. However, it is unfortunate that the major part of this material cannot be assigned to its cultural provenience, since it was collected before stratigraphy had been developed.

The Necropolis of Ancón reflects the later periods on the Central Coast after the agricultural pattern had become established; likewise, those native plants of the coast that had potential industrial uses had already found their place in the cultural framework. Added to these established economic plants were a number of both cultivated and wild highland importations, such as one would expect to find as the result of the diffusion of the Tiahuanacoid influence from that area. Some highland introductions should perhaps be attributed to the later Inca influence. Uhle's discovery of Inca graves at a location beyond the Necropolis and the occurrence of Inca types of pottery among the ceramics recovered by Reiss and Stübel attest to this later integration.

The cultivated plants found at Ancón which were already common to coastal Peru include the following: maize, common beans, lima beans, peanuts, sweet potato, manioc, achira, chile pepper, Hubbard squash, crookneck squash and gourds (possibly used for food). Species which were probably derived from the highlands are quinoa, ulluco and possibly lupine. Several wild species, some of which may actually have been grown for their fruits, were found: avocado, guava, lucuma, algarroba, pacae and *Campomanesia lineatifolia.* The following species with edible fruits have been mentioned as occurring at Ancón either as actual plant remains or in artistic forms: *Aristotelia Macqui, Caryocar amygdaliferum, Achras Zapota* and *Annona Cherimolia.*

Plants which supplied material for the manufacture of artifacts and those which presumably were used for their medicinal qualities were:

Industrial Plants

Fiber Plants

Typha angustifolia (cat-tail)
Carludovica palmata
Furcraea sp.
Ceiba pentandra (possibly)
Bactris ciliata
Guilielma speciosa
Gossypium barbadense (cotton)
Species of *Tillandsia*
Sporobolus sp.

Dye Plants

Dicliptera Hookeriana
Dicliptera peruviana
Relbunium nitidum
Bixa Orellana
Piper lineatum
Roupala ferruginea
Caesalpinia Paipai
Caesalpinia spinosa
Arrabidaea Chica
Lafoensia punicaefolia
Miconia chrysophilla

Woods

Pineda incana
Porlieria hygrometra
Pavonia paniculata
Godoya obovata

Beads and Ornaments

Thevetia peruviana
Achras Zapota
Phytelephas macrocarpa
Nectandra reticulata
Ormosia sp.

Containers

Lagenaria siceraria (gourd)
Crescentia Cujete (calabash tree)

Drug Plants

Erythroxylon Coca
Erythroxylon novogranatense
Andira inermis
Mucuna (probably *M. inflexa*)
Piper asperifolium
Ilex sp. (possibly)

Miscellaneous

Sapindus Saponaria
Myroxylon sp.
Acacia macrocantha (?)

The extensive ruins of the city of Pachacamac are located near the mouth of the Lurín Valley some six hundred yards from the ocean. Within the city proper, there are two mounds, each representing a temple—one, the Temple of Pachacamac, the other, the Inca Temple of the Sun. The Temple of Pachacamac is the older of the two. Uhle (1903), who surveyed the site in 1896 and carefully excavated parts of it, found the remains of walls of the "Old Temple" beneath those of the present structure. Additional evidence of the age of the site was found in the crowded cemetery at the northern base of the Temple.

Pottery styles of the Coastal Tiahuanaco (or possibly an earlier period) were recovered here.

The Temple of the Sun and the cemetery located on its southeast terrace are undoubtedly of Inca origin. Uhle (*ibid.*) found that the terrace burials contained Inca textiles and pottery, as well as vegetal remains definitely associated with the highlands. More recently Tello made extensive explorations at Pachacamac. It was at this latter time that Strong (1943), in an attempt to establish a pottery sequence, made two stratigraphic cuts in one of the rubbish heaps at the base of the Temple of the Sun. A large amount of material which contained plant remains, distinctive both in number and representative species, was recovered. The plants found in the Playa Grande levels in the refuse heap include cotton, gourds, lucuma and specimens of wood and caña brava. The list of species from the later levels, which Strong designates as Inca-associated and Inca, respectively, is more complete (Towle, 1948). The common cultigens are included: maize, achira, peanuts, common bean, lima bean, cotton, chile pepper, crookneck squash, gourds, and two highland tubers, oca and the white potato.

Wild plants which may well have been used for food are represented by avocado, pacae, *Canavalia,* lucuma, *Campomanesia lineatifolia,* and ciruela del fraile (*Bunchosia,* sp., probably *B. armeniaca*). Additional species include *Phragmites communis, Calamagrostis* sp., *Scirpus* sp., *Mucuna* sp., *Sapindus Saponaria,* and *Erythrina* sp. (Towle, 1948).

To these lists may be added the following, whose cultural associations are unknown: a species of bean, manioc, sweet potato and Hubbard squash, and also *Guilielma insignis, Nectandra* sp., *Caesalpinia Paipai, Erythrina* sp. (possibly), *Erythroxylon novogranatense.*

All of the above species have been reported from other coastal sites, including the Ancón Necropolis. However, Uhle observes that although some of the more common foods found on the coast, such as manioc and the sweet potato, are lacking in the Inca graves on the terrace of the Temple of the Sun, both black and white *chuño,* a food similar to *chuño* made of oca and large amounts of quinoa seeds (all staple foods of the highlands), accompanied the burials. In addition, coca, chile peppers, maize and two varieties of beans were often included.

All the species represented among the plant remains collected by Stumer in levels of the Fusion Epoch at sites in the Rimac and Chillón Valleys and examined by the writer are common to the Central Coast during and after the Late Formative. These plants include both wild and cultivated species (Towle, 1958).

Besides the vegetal remains from the Central Coast Sites discussed above, there is additional material that has been collected from other localities in the Chancay, Ancón, Chillón and Rimac Valleys. Much of this can be culturally assigned only in general terms, since it was collected without the benefit of stratigraphy. However, it contains no species not included in the list of cultivated and wild plants mentioned above. The lack of information on the cultural provenience of this material makes its chronology incomplete.

THE SOUTH COAST

The South Coast is limited by the Rio Mala on the north and by the Rio Lomas on the south. From the available data it appears that this region may be subdivided into two sub-areas, the Cañete and Chincha Valleys comprising one, the Pisco, Ica and Rio Grande de Nazca Valleys the other. Although much research has been conducted in the South Coast region, the most satisfactory information that we have pertains to the latter sub-area (Tello, 1929; Uhle, 1914; Kroeber and Strong, 1924, a, b; Gayton and Kroeber, 1927; Strong, 1954, 1957), from which the familiar, highly decorative ceramics and textiles to be found in museums were obtained. As yet it is difficult to determine the relative importance of the small valleys to the south of the Rio Grande because of the lack of archaeological information. This region is bleak and exceedingly rugged (Stumer, 1954d). Kroeber (1944) notes that within these valleys there is scattered evidence of Nazca influence, as well as of the later Ica-Chincha culture.

The South Coast contains numerous sites of a variety of types, ranging from extensive shell mounds on the coast, which belong to the earlier developmental period, to the complex constructions and settlement units which are characteristic of the Fusion and subsequent Epochs.

Although a tentative cultural sequence had been established for the South Coast, some confusion exists in the developmental picture which can only be clarified by extensive survey and stratigraphic work. An expedition to collect such needed data was recently conducted by Strong (1954; 1957) in the Ica and Rio Grande Valleys. His purpose was three-fold: to collect information for a dependable ceramic sequence, to trace the cultural distribution in the valleys, and to study the settlement pattern of each. The results of his findings have not as yet been published, but the preliminary reports (*ibid.*) give promise of a clearer picture of South Coast cultural history.

The ethnobotany of the region is derived largely from sites located in the Pisco, Ica and Rio Grande Valleys and on the peninsular of Paracas.

Epoch of Incipient Agriculture

South of the mouth of the Rio Grande on the Bays of Chavinia and Lomas there are extensive shell mounds. Those at the latter location were explored by Strong only to the extent of demonstrating that pottery was lacking in the lower levels and that some plant material and remains of textiles were present. The occurrence of these two cultural elements without the association of pottery is reminiscent of the lower levels of Huaca Prieta. More recent explorations in the Ica shell mounds have identified part of that site as belonging to this early epoch.

Formative Epoch

Evidence of the Early Ancón Period of the Central Coast has been observed at shell heaps located on the coastal sand dunes between the Rio Lurín and the Rio Chilca (Strong and Willey, 1943), thus extending this early manifestation of the Formative into the southern coastal division. Extensive ruins are also reported from the Mala Valley (*ibid.*). However, the entire region remains to be explored scientifically.

As mentioned previously, the most satisfactory archaeological picture for the South Coast has been obtained from the Pisco, Ica and Rio Grande area, but here there has been some lack of agreement as to the sequence of the periods involved. The resulting confusion is on the way toward clarification with the ceramic studies conducted by Strong in connection with his explorations in this general region. He recognizes three periods as constituting the Formative: Early Paracas, Late Paracas and Proto-Nazca (Strong, 1957). Recently a still earlier, pure Chavinoid culture has been recognized at Cerrillos in the Ica Valley. This culture is contemporaneous with those of Cupisnique and Early Ancón. The center for the dispersal of the Paracas culture has generally been considered to be the site of Paracas located on the hill of Cerro Colorado and the adjacent plain on the Peninsula of Paracas. This peninsula is situated between the Pisco and Ica Rivers, and like the adjoining mainland, is virtually a desert, supporting little or no vegetation. Nevertheless, what appear to be two well-defined and yet apparently related cultures have been described at this site. However, Strong maintains that these are two phases of a single culture; in fact, there is some indication of contemporaneity between them.

Early and Late Paracas. On the slopes of the Cerro, a cemetery designated as "Paracas Cavernas" (Tello, 1929; Carrion, 1949) typifies the earlier of these two cultural phases. Here numerous burials and grave equipment were placed in subterranean tombs cut in the soft rock. The culture contains many elements also found in the site of Paracas Necropolis as well as elsewhere in the Ica-Nazca region on the mainland. The textiles that characterize the Early Paracas culture are relatively simple, whereas the heavy, incised pottery is distinguished by a boldness of design which suggests the art of the Chavín Horizon. Although of possibly tenuous kinship, the early culture of the Peninsula is "rooted in the Chavín tradition" (Strong and Evans, 1952, p. 244).

On the lower slopes of Cerro Colorado, another cemetery was discovered. The "Paracas Necropolis," as it is called, typifies the second and later phase (Late Paracas) of the culture occurring at Paracas (Tello, 1929; Strong, 1957). The crowded burials of this cemetery were placed in underground houses with thick walls of stone and adobe. Although the cultural kinship between the two sites is apparent, both the textiles and pottery that were recovered are in sharp contrast. The simple textiles of Cavernas were replaced in the later Paracas phase by fabrics which show a diversity of techniques and a richness of design that recalls the ceramic art of Nazca. It is noteworthy, moreover, that the simple, polished, undecorated ware of the Necropolis is in marked contrast to the incised, polychrome pottery of Cavernas.

Both Early and Late Paracas culture have also been recognized in the Ica-Nazca region. Strong and Willey (1943, p. 22), as a result of their survey of the Ica Valley, conclude, at least tentatively, that the two Paracas cultures represent either two contemporaneous cultures on the South Coast or possibly two stages of the same culture which later developed into the Early Nazca. Strong's more recent studies appear to confirm this latter hypothesis (Strong, 1957).

As yet no settlements attributable to the Early Paracas period have been recorded on the mainland. Evidence of the Late Paracas culture is, however, widespread. Burials at Ocucaje in the Ica Valley and house structures with accompanying refuse heaps at Cahuachi in the Nazca reflect its presence.

In the description of the Paracas cemeteries on the Paracas Peninsula, mention was made of the lack of sizable associated population centers. This has led to speculation as to the possible communities to which these large burial areas belonged. Either the cemeteries are the remaining evidence of a once-flourishing local community depending largely upon agriculture for its subsistence or the mummy bundles were brought to the peninsula from such a community located elsewhere. Arguments have been presented for both of these possible explanations. The cultural links existing between Paracas and the Ica-Nazca Valleys seem to present some argument in favor of the Ica-Nazca region. In fact, it has been suggested that the more important inhabitants of the city were prepared for burial at Cahuachi and later taken to the cemeteries of Ocucaje or to those on the Paracas Peninsula. This hypothesis remains to be proved.

The most abundant as well as the best documented evidence for the use of plants by the people of the South Coast during the Formative Epoch has come from the sites of Paracas and of Cahuachi. The majority of the remains from the former area were placed within the mummy bundles as artifacts of plant origin or as funerary offerings of food; the greater part of the collection made at Cahuachi during the excavations conducted by Strong were from seven strata-cuts in the living areas. The plant remains from Paracas included both wild and cultivated species. From the older Cavernas site, the domesticated species consisted of maize, cotton, gourds, pepper, and the lima bean (Mangelsdorf, 1942; Yacovleff and Herrera, 1934), whereas the plants furnishing materials for making artifacts have been identified as a sedge (*Scirpus* sp.), reeds and a palm (Yacovleff and Muelle, 1932, pp. 133–34; Yacovleff and Herrera, 1934, pp. 263–64). Furthermore, coca was recovered from a tomb at this site. A larger number of species of both cultivated and wild plants have come from the Necropolis. With the exception of the lima bean, the four domesticated plants found at Cavernas were also discovered at the Necropolis site, and in addition several others make their appearance. These latter include the peanut, jicama, manioc, the common bean and the sweet potato (Towle, 1952b; Carrion, 1949). The palm is not recorded for the Necropolis, but sedges and reeds are mentioned, and the list of wild plants is further increased to include an alga, the cat-tail, caña brava and a fiber plant, *Furcraea* sp. (Towle, *ibid.;* Yacovleff and Muelle, 1934; Yacovleff and Herrera, 1934, pp. 263–64).

The literature refers to plant remains recovered from Paracas without indicating the exact location where they were found. Some of these plants have already been mentioned as occurring at either Cavernas or Necropolis and in some instances at both sites. Others not represented in these lists include six species of wild plants. Two of these (lucuma and *Canavalia*) may possibly have been cultivated for food and the other four (calabash, a willow, *Baccharis* and *Relbunium*) used in industry. There are three references to pottery vessels found at Paracas that resemble fruits. These are the domesticated squash, the pepino and the lucuma.

Although the plant remains from Cahuachi were largely obtained from the seven strata-cuts, other specimens came from two burial areas, several burials therein, the summit of Mound No. 4 and the Great Temple site (Towle, 1956). Those remains from the Late Paracas site of Ocucaje consisted of specimens from three burials and a small collection from the subsurface layer (*ibid.*). Of the twenty genera represented, ten species were of cultivated plants, three that may have been cultivated, and seven of wild species. Although many of the same species of cultigens were found at both Paracas and at Cahuachi, two of those frequently encountered at the former site are lacking at the latter, namely, jicama and the pepper. However, the provenience in which the plant remains of each site were located may explain this. Specimens of the more common food plants were intentionally placed with the burials at the Cavernas and Necropolis at Paracas, whereas the remains at Cahuachi were in the refuse only by chance.

Proto-Nazca Period. This is the final period in the Formative Epoch in South Coast chronology. Evidence of this culture has been found only in refuse deposits at Cahuachi. As demonstrated by the pottery recovered, Proto-Nazca is transitional between the Late Paracas and Early Nazca cultures. Among the ceramic styles that characterize the period are those which bear geometric and naturalistic designs that include depictions of "beans and other plants" (Strong, 1957). The plant species represented in levels of this period at Cahuachi were also included in those of the Late Paracas and Early Nazca (Towle, 1956).

Classic Epoch

Nazca Periods. The peak in art and master craftsmanship, represented by the Mochica Period on the North Coast, was also achieved by the Nazca culture on the South. This culture had its center in the Nazca Valley whence it spread to that of the Ica some twenty-five kilometers to the north. It is best known for its elaborate textiles and polychrome pottery. Based upon stylistic changes occurring in the pottery, the Nazca Period has been subdivided into Early, Middle and Late Nazca.

The majority of the Nazca sites are located in the upper Nazca Valley and these are mostly cemeteries that lie on the valley margins. The largest and most famous of these is Cahuachi, from which superb textiles have been obtained. In fact, it seems that this site was the Nazca capital throughout the Classic Epoch (Strong, 1954). Here is to be found the largest temple and pyramid complex of the Middle Nazca Period. Few sites, if any, compare with it either in size or complexity. As a matter of fact, there is little evidence of large organized communities. This apparent lack points to small settlements of farmers with well-developed artistic and technical ability.

By the close of the Classic Epoch, all the major cultivated food plants of Coastal Peru had been domesticated and had become an accepted part of the agricultural pattern. Many of these plants have been found as actual remains, while others are reproduced in the decorative and plastic arts so characteristic of Nazca culture. The species of domesticated plants associated with Early, Middle and Late Nazca include manioc, achira, aji, jicama, the common bean, the lima bean, peanut, maize, cotton, gourds, sweet potato and both Hubbard and crookneck squashes. Supplementing this list of edible species, there should be added three wild species which may have been cultivated: *Canavalia,* a guava and an avocado. Coca was also used, as well as a number of plants in industry. Other species that were known and possibly used in different ways include algarroba, an alga, a cactus, pacae, *Campomanesia lineatifolia,* soapberry, reeds and sedges.

There are evidences of agricultural practices within the Ica-Nazca area in the form of canals and irrigation ditches, some of which may safely be placed in the Formative Epoch. The amount of arable land in the Ica and Rio Grande Valleys is limited and therefore could not have supported a large population (Kroeber and Strong, 1924b, p. 24). The valleys of the South Coast, although large, are not as well watered as those of the North; in fact, some of the rivers are semi-permanent, becoming dry before they reach the sea (Bennett, 1946b, p. 92). In the Nazca Valley, in addition to surface irrigation there are the remains of underground aqueducts which were built to bring the seepage water from the tributaries of the upper valley. These constructions, several of which are used at the present time, are two hundred to two thousand feet long, stone-lined and braced with logs (Kidder, A. II, 1947 in conversation). It is thought (Kroeber and Strong, 1924b, p. 122) that the water supply in the lower Ica Valley was greater in pre-Inca times than it is today, as there are evidences of cultivation both above and below Ocucaje which, at the present time, lacks accessible water. This condition has been remedied by tapping the river above the city of Ica by a canal attributable to the Incas, which follows along the eastern edge of the valley.

Epoch of Fusion

There are numerous sites in the Nazca Valley in which Tiahuanaco and related Tiahuanaco styles occur. These sites include large cities with great stone enclosures containing thousands of rooms. The site of Huaca del Loro represents such a community. The ceramics recovered here show characteristics of Late Nazca as well as a strain of Tiahuanaco which may have come from the Tiahuanaco site of Huari to the northeast (Strong, 1957). Plant remains found at this location include maize, achira, peanut, lima bean, cotton, aji, crookneck squash, manioc, and the common gourd. Still others recovered were *Canavalia,* pacae, algarroba, soapberry, lucuma and *Campomanesia lineatifolia* (Towle, 1956).

The "Pacheco" culture also reflects this epoch of Tiahuanacoid influence, its pottery incorporating both Nazca and Tiahuanaco elements. However, our knowledge of this manifestation of the Epoch of Fusion is limited; its extent in time and space are yet to be determined (Strong, 1957).

The actual plant remains and the plant representations in art which have been found in sites of this Epoch include the group of domesticated and wild plants of the earlier Nazca Periods. There are, however, several additions to the list. These plants have been identified in the Tiahuanacoid designs on "Pacheco" pottery and constitute five species grown in the highlands.

The species include a lupine, three root crops (añu, ullucu and oca) and the potato. The latter is represented in the art of the North Coast during Mochica times, but the writer has not found references to its occurrence on the South Coast before that of the "Pacheco" culture. These species may have been actually introduced into the agriculture of the region, or the textiles and ceramics which portray the design may themselves have been brought from the highlands.

Subsequent Epochs

From the available evidence, the Epoch of Kingdoms and Confederacies and the Imperial Epoch saw little change in either the ethnobotany or the agricultural methods of the South Coast. However, the increased complexity of the socio-political system during these final centuries of South Coast chronology doubtless was reflected in the subsistence economy. Such changes as might appear in a wider dispersal of crop plants and agricultural techniques are yet to be found.

THE HIGHLANDS

North Highlands. The Andean chain in the North Highlands extends from the Ecuadorian border to Huánuco. Although much of this great area is inaccessible, a number of archaeological sites are known, particularly in the valleys of the Callejón de Huaylas and Marañon. The cultures represented at these locations have made it possible to construct a chronology for the North Highlands, which, although not complete, presents a sequence extending from the Formative into the Epoch of Kingdoms and Confederacies. Among the best known of the sites are the classic Chavín de Huántar located in the Marañon Valley and Recuay in the valley of Callejón de Huaylas.

With the exception of a few scattered sites of the Epoch of Pre-Agriculture, the earliest culture yet discovered in the North Highlands, or in the entire highland region of the Central Andes for that matter, is that of Chavín, which was first noted at the site of Chavín de Huántar. The distinctive character of the Chavín culture is reflected in the stone carvings, ceramics and architecture, not only of the North Highlands but on the coast as well, although there are a number of basic differences between the two. Our knowledge of coastal Chavín is more extensive than that of the highland manifestation.

Although the center of dispersal of Chavín is in the North Highlands, there is no evidence to support the belief that the site of Chavín de Huántar was its genesis or the location from which it was diffused. It is of particular interest to take into account the fact that the appearance of Chavín elements on the coast was accompanied by the important domesticated plants, maize, peanuts, beans, and possibly manioc.

The Formative Period of Recuay is well represented at the site of Recuay, from which the period derives its name. The ceramic art, if not refined, is well developed. Two types of pottery are present; one, which is thought to be the older, resembles gourd forms, while the other includes molded forms crudely executed and varied in shape. Some of these latter are copies of fruits and are reminiscent of the type of pottery found in Mochica and Chimu cultures. Tello has suggested that the tripod-form, common in Recuay pottery, probably was inspired by the root clusters of the *arracacha* (Kroeber, 1944, p. 93).

There is little ethnobotanical information on this area. The climate of the highlands is not favorable to the preservation of perishable materials and consequently there are few references to the occurrence of plant remains. Seeds of *Erythrina* sp. have been found at Recuay (Yacovleff and Herrera, 1935, p. 43) and the pottery vessels made in the form of plant parts that have been found there give some indication as to what species were used.

Central Highlands. The Central Andean highlands extend approximately from Huánuco to beyond Cuzco. Although this section of rugged terrain contains numerous sites, few have been investigated, and our knowledge of the cultural history of the area is derived largely from locations in the region around Cuzco and from archaeological studies at the Inca capital itself, all of which represent the Inca Period.

The culture found at the site of Chanapata west of Cuzco has been placed in the Formative Epoch. Here were found well-developed pottery, houses that may have been semi-underground, and stone-faced terraces that resemble those discovered at Chiripa on the south shore of Lake Titicaca. In both instances, these may be the remains of agricultural terraces, although Rowe (1946, p. 198) doubts that the Chanapata terraces were built for that purpose.

The extensive Inca ruins at and near Cuzco are well known and many of them have been carefully excavated and described. However, there is still little information on the origin and early development of Inca culture. From their traditional history, the Incas had occupied the site of Cuzco for at least three hundred years before the coming of the Spanish. Rowe, as a result of his comprehensive study of the chroniclers of the Spanish Conquest and Colonial periods, recognizes an early and a late stage in the history of Inca culture, to which he has assigned the dates of A.D. 1200–1438 and A.D. 1438–1532 respectively. Accordingly, the Early Inca culture was existing during the Epoch of Kingdoms and Confederacies.

Late Inca, a period of approximately one hundred years, encompasses the preliminary conquests and the ultimate formation of the Inca Empire. There was little change in the general cultural pattern of the conquered peoples, and what changes were made were for the benefit of the Inca state as a whole. The chroniclers have left detailed accounts of the elaborate economic system, which was based upon intensive agriculture. This was perfected in order to meet the demands of the growing population as well as an elaborate political and religious structure. These historical documents give imposing lists of cultivated species and also mention the wild plants that were part of Inca ethnobotany. The wide range of cultivated plants was due to the varying altitudes and climatic conditions. Tropical, semi-tropical and temperate conditions in the highland valleys made possible the domestication of a number of species, most of which were also cultivated in the coastal valleys where similar growing conditions existed. The cooler table-lands also had their characteristic species, such as quinoa, the potato, and the endemic tubers, añu, oca and ullucu. In fact, each successive altitudinal level up to 13,000 feet, the limit of agriculture, not only had its own natural flora but its cultigens as well.

There was much interchange of plant species throughout the Empire, partly the result of the general agricultural plan and partly due to the movements of people from one area to another. Many wild and cultivated plants have been identified from remains recovered from Inca levels on the coast, but such material is rarely found in the highlands. Rowe (1946, p. 243) mentions the occurrence of specimens of basketry and matting in tombs near Cuzco. Floral elements in Inca designs have further added to our knowledge of the plants used. Many of these have been identified, and it is interesting to find that some of the species in use today were similarly utilized in pre-Spanish times (Yacovleff and Herrera, 1934, 1935; Herrera, 1923).

In the highlands, terracing and irrigation were perfected agricultural practices and the many impressive remains have been the subject of much discussion and speculation. Again we must turn to the chroniclers for information on the tools that were used—the foot-plow, hoe, clod-breaker and scraper. There are also detailed accounts of the ceremonies accompanying planting and harvesting. Within the conquered coastal regions, agricultural practices, although remaining unchanged, were often enlarged and perfected.

South Highlands. This region is an extensive plateau which consists of the Puno of Peru and the altiplano of Bolivia, the latter containing most of Lake Titicaca. With the exception of the famous site of Tiahuanaco, archaeological research has largely been restricted to the immediate Titicaca area.

The earliest culture that has been isolated for the region is found at Chiripa, a site on the south shore of Lake Titicaca, which is apparently pre-Tiahuanaco (Kidder, 1956) and which has been placed in the Formative Epoch. The dwellings were low-walled, rectangular structures placed in a circle with their front corners touching. The walls were double, the intervening spaces having been used for storage. In one such bin a large amount of quinoa was discovered (Bennett and Bird, 1949, p. 142). Charred remains of potatoes were also recovered from the site, as were specimens of charred cordage and fragments of basketry (Kidder, 1956).

Following the Chiripa Period in the South Highland cultural sequence are the Early Tiahuanaco and the Classic Tiahuanaco Periods of the Classic Epoch.

Classic Tiahuanaco in its purest form is best demonstrated at the site of Tiahuanaco. This extensive area, located thirty-four miles south of Lake Titicaca. is one of the best known in the Andean region and has been the subject of much detailed study. It is difficult to reconstruct the general culture of Tiahuanaco. Although not yet known, it seems that the economy was based, at least in part, upon agriculture, the common highland cultigens furnishing the staple plant foods. Potatoes, oca and quinoa were grown in all probability in the limited farming area of the region, as they are today, while other foods such as ullucu, añu and lupine, if not cultivated, were probably obtained through trade.

The climate of all the Andean highland area is not conducive to the preservation of vegetal remains and we have only scant evidence of the plants—either wild or cultivated—that were used. However, we assume that the more common native crop plants grown in the intermountain valleys today were probably cultivated by the prehistoric inhabitants. Likewise, those cultigens which augment the meat diet of the herdsmen of the plateaus were probably known at an earlier date and were grown on the sparse farm land there. The finding of charred potatoes and quinoa at the site of Chiripa gives credence to this assumption.

As in the case of the plant remains, there are few clues upon which we can reconstruct the agricultural methods that were employed. There are remains of agricultural terraces on the lower slopes of the sides of the hills that rise from the river flats of the deep valleys, some of which are used at the present time. Also there are terraces at Chanapata and at Chiripa which may have been used in the agricultural practices of their inhabitants. The history and development of cultivated plants and of agriculture in the highlands is highly speculative; in fact, the ethnobotany of the cultures to be found there is largely hypothetical.

Bennett and Bird (1949, p. 157) state that some agricultural terraces found in the South Highlands have been identified as belonging to this Mastercraftsman Period, which spans not only the Classic Epoch but also part of the Fusion Epoch. These constructions may possibly be the same referred to by Bennett (1946a, p. 21) as being located in Bolivia and belonging to the Tiahuanaco periods.

SUMMARY

The ethnobotanical data and the evidence regarding the related agricultural practices in the several cultural-geographical subdivisions as presented in the preceding sections do not reflect the evolutionary trends that existed in the region as a whole. These data are consequently summarized in the following pages in order to make clear the nature and extent of such trends in the cultural development of the Central Andes.

Sites of the *Epoch of Pre-Agriculture* which have been discovered in Peru give little or no information as to the use of plants by the hunter-gatherer inhabitants. It stands to reason, however, that plants were used as a source of food and to furnish materials for the manufacture of artifacts. The full extent to which the available food and industrial species were employed is yet to be defined. Certainly by the end of the Epoch of Pre-Agriculture the ethnobotanical aspects of this early culture must have already gone through a long, developmental period of experimentation.

The first tangible evidence of the growing of domesticated plants marks the beginning of the *Epoch of Incipient Agriculture,* an era now divided into a Pre-Ceramic and an Initial Ceramic stage (Collier: 1960 [MS]). Only a few sites have as yet been discovered that represent this early manifestation, which was to develop into the most highly organized agricultural system in the New World in prehistoric times. These sites are the remains of small villages of subterranean and semi-subterranean houses located on the coast, the inhabitants of which depended upon an economy of fishing, gathering and horticulture. The remains of a number of species representing both cultivated and wild forms have been found, the latter group furnishing both food and industrial materials. At first it was a pre-ceramic culture; the people were dependent upon baskets and gourd containers, both of which have been recovered. Mats were made and twining and weaving were practiced, a native cotton and various other fibrous plants supplying the necessary materials. The cultivated species are limited, the absence of maize being the most notable feature. For this reason the earlier periods of the Epoch of Incipient Agriculture are often described as representing a pre-maize culture.

The first pre-ceramic site attributable to this epoch, and the one from which the most complete information has been obtained, is Huaca Prieta in the Chicama

Valley on the North Coast. A large number of vegetal remains were recovered here. The species that undoubtedly were domesticated include two types of squash, the lima bean and the common gourd.

The North Coast culture of Early Guañape is assigned to the Initial Ceramic stage of this Epoch. This manifestation is outstanding because it introduces pottery for the first time into Central Andean chronology. Except for cotton and gourds, no vegetal remains have so far been reported for Early Guañape levels in the Virú Valley, while the species representative of the period at Huaca Prieta are yet to be described.

A site which appears to belong to the Initial Ceramic stage of the Epoch of Incipient Agriculture is the Aspero Midden on the Central Coast. Because of the absence of pottery at the site itself, it has been considered "pre-ceramic." However, the finding of a cache of maize beneath the Early Ancón level would place it within the agricultural period.

The vegetal remains from levels assigned to the Epoch of Incipient Agriculture consist of the following:

Cultivated

Squash (*Cucurbita moschata*)
Squash (*C. ficifolia*)
Gourd
Lima beans (*Phaseolus lunatus*)

Possibly cultivated

Chili pepper
Cotton
Achira
Lucuma
Ciruela del fraile (*Bunchosia armeniaca*)

Wild Species

Cat-tail
Sedges (*Scirpus, Cyperus*)
Canes
Algarroba
Bast Fiber
Pounded bark–cloth

Evidence of the type of agriculture practiced or the tools that were used has not been found. It appears, however, that the cultivated plants were grown in small garden plots on the marginal lands in the lower river valleys.

The *Formative Epoch* which followed the Epoch of Incipient Agriculture was one of long duration and progressive cultural activity. During its earlier periods there was an increase in the coastal population; the communities remained small in size, with some larger centers being built about platformed temples in the valleys south of the Virú.

The economy continued to be based upon fishing and agriculture, the latter reflecting advances in methods and in the number of cultivated species. In the Cupisnique levels at the site of Huaca Prieta in the Chicama Valley, maize, peanuts, and the common bean are added to the cultivated plants of the earlier Epoch. These additions are of particular interest, since they make their appearance with the elements of Chavín influence, a definite highland manifestation. Peanuts and manioc, which have been reported from contemporaneous Early Ancón levels on the Central Coast, constitute a cultivation complex that is associated with the warm temperate valleys of the Amazon drainage system. Their appearance on the coast brings to mind Tello's theory (1930) that cultigens characteristic of the inter-Andean valleys were possibly carried thence by traders or migrating groups down to the coastal valleys. Coca is also added to the list of plants of the former Epoch. In addition, there is evidence of the use of three species that may have been cultivated, including two varieties of pacae, avocado and the edible fruit, caigua. With the increased production of manufactured objects, the utilization of industrial plants must also have increased proportionately.

The few highland sites assigned to the first part of the Formative have contributed little to our knowledge of the agricultural economy upon which they were undoubtedly, at least in part, dependent. The only tangible evidence has been found at Chiripa in the South Highlands, where specimens of quinoa and the potato appear.

The later periods of the Formative witnessed a continued growth in population with settlements of varying size and complexity in the river valleys away from the shore. Agriculture had now become established as the major economy, with the definite appearance of irrigation canals and their constantly increasing use.

Certain species of cultivated plants recovered from sites in coastal areas attributed to the early periods of the Formative have not been reported from some of the later ones. This fact, however, should not be construed as indicating a discontinuance of their use since, with the exception of *C. ficifolia* and possibly caigua, all are found in the subsequent Classic Epoch. Three species, however, are added to the list of the cultigens grown during the early part of the Formative. These are pepino (*Solanum muricatum*), sweet potato and jicama. The "wild plum," (*Campomanesia*) also appears.

The wild species used for food and in industry during the earlier part of the Epoch continue to appear among the collections of vegetal remains from the later period.

The plant species associated with the cultures of the Formative Epoch are:

Cultivated

Chile pepper
Common beans (*Phaseolus vulgaris*)
Cotton
Gourd
Jicama
Lima beans (*Phaseolus lunatus*)
Maize
Manioc
Peanut
Pepino (*Solanum muricatum*)
Potato
Quinoa
Squash (*Cucurbita moschata*)
Squash (*C. ficifolia*)
Sweet potato

Possibly cultivated

Achira
Avocado
Caigua
Campomanesia
Coca
Guava
Pacae

With the exception of certain artificial ditches found in the upper Virú Valley which have been definitely assigned to the Puerto Moorin period, there seems to be no conclusive evidence for the use of irrigation on the coast before the Gallinazo I Period of the North Coast and the Playa Grande of the Central Coast (Willey, 1953, p. 362; Larco, 1946, p. 156). The date of its inception on the South Coast still remains obscure. During Gallinazo times on the North Coast, irrigation developed into the large-scale canal systems that supplied water to the cultivated fields of the upper valley areas.

The increasing occurrence of stone hoes and digging sticks in Formative sites is further evidence of greater agricultural activity.

There is little available information on the agricultural methods used in the highlands at this time. Terraces at the sites of Chanapata and Chiripa are thought by some archaeologists to have been part of the agricultural system, but this interpretation must remain open to question, at least for the present.

The subsequent *Classic Epoch* on the North and South Coast is characterized by the growth of regional cultures in which the technologies of the previous

Formative Epoch reached the peak of their development. There does not appear to have been a similar florescence on the Central Coast, although there was a continued growth and general cultural advance. Settlements during the Epoch were numerous, and ranged from fair-sized villages to large population-centers situated in the major coastal valleys. Subsistence was almost completely dependent upon intensive agriculture; the population maximum which was achieved at this time demonstrates the interaction between food supply and population increase.

The number of domesticated species and wild food plants utilized during this Classic Epoch is impressive. In fact, it is noteworthy that no additional cultigens appear until the time of the Spanish Conquest. The wild food plants that were first observed during this Epoch probably had been part of the vegetable diet in previous periods, but no remains of the plants or molded forms have been found or reported. Both Mochica and Nazca pottery, however, are noted for their ceramic reproductions and pottery designs of cultivated or wild species.

The species utilized in the coastal area include:

Cultivated

Common beans (*Phaseolus vulgaris*)
Chile pepper
Coca
Cotton
Gourd
Jicama
Lima beans (*Phaseolus lunatus*)
Llacon
Maize
Manioc
Peanut
Pepino
Potato
Squash (*Cucurbita maxima*)
Squash (*C. moschata*)
Sweet potato

It will be noted that the squash, *C. ficifolia*, is not in this list. This species has not been reported later than the Cupisnique levels at Huaca Prieta. However, the Hubbard squash, *C. maxima*, is found for the first time.

Possibly cultivated

Achira
Algarroba
Avocado

Campomanesia
Canavalia
Caigua
Cherimoya
Ciruela del fraile
Guanabana
Granadilla
Guava
Lucuma
Pacae
Papaya
Pineapple
Tumbo
Tuna

The cultigens associated with highland agriculture comprise:

Añu
Cañihua
Lupine
Oca
Potato
Quinoa
Ullucu

The high level of agricultural development in the coastal areas is reflected in the intricate valley-wide irrigation systems. The canals and aqueducts are often of impressive size. Agricultural tools in the form of hoes and digging sticks are commonly found, and it has been suggested that guano fertilizer was utilized (Bennett and Bird, 1949, p. 158). It is indeed possible that nearly all the agricultural techniques associated with the prehistoric cultures of the Central Andes had been developed and were in use by the end of the Classic Epoch.

Although little information is available on agricultural methods as practiced in the highland regions, the evidence we have seems to indicate an active agricultural economy in that area. Terraces have been found in the South Highlands and assigned to the Classic Epoch (the Mastercraftsman Period of Bennett and Bird, 1949, p. 157), but their genesis and development is as yet unknown.

The outstanding feature of the *Epoch of Fusion* is the widespread occurrence of the Tiahuanaco Horizon complex, which serves to bind the cultures of the coastal and highland regions into a common relationship. Remains of cultivated plants from the coast represent species already familiar from the previous epochs. A wider distribution of certain of the highland cultigens is apparent, which would be expected in light of the general dispersal of Tiahuanaco culture. In contrast to the firmly established food plants, plants of industrial and medicinal use increased perceptibly during the Epoch of Fusion and also the later Epoch of Kingdoms and Confederacies. These species, largely identified from among the vegetable remains recovered at the site of Ancón, furnished fibers, woods, dyes, medicines and drugs. No other Central Andean site has contributed so large a number of both wild and cultivated forms. Intensive agriculture continued to be practiced in the coastal valleys.

The Epoch of Fusion lasted for about two hundred years. In coastal Peru its end was marked by the general breakdown of whatever political structure was in existence and the formation of four local kingdoms, dominated by a hierarchy of hereditary nobility. This change in political organization ushered in the *Epoch of Kingdoms and Confederacies.* During the ensuing centuries, the population, which remained large, lived in organized urban communities. Trade was further expanded and coastwise contact by means of the sea was established between Peru and Ecuador.

The ethnobotanical picture resembles that of the Fusion Epoch, although it is better demonstrated by the recovered plant remains. Greater trade activity must have increased the number and distribution of important species. The agricultural practices of the previous epochs continued to be followed, as well as the use of plots of *hoyas* (*pukios*) on the North Coast.

During this Epoch, Inca culture, with its nucleus at Cuzco, was on the ascendant. Although there is little information on its agricultural economy, it seems reasonable to assume that the agricultural pattern so well developed in the Late Inca (Imperial Epoch) was having its foundations established during this earlier stage. The fact that terracing has been recognized in previous epochs and was to become such an important agricultural practice during the Inca regime makes its extensive use during the Epoch of Kingdoms and Confederacies seem assured. The same is probably true of irrigation. The difficulty of dating such constructions is a drawback in tracing their development and utilization.

The *Imperial Epoch* witnessed the expansion of Inca domination and the establishment of the Inca Empire. Because of the records left by the chroniclers of the Spanish Conquest and the following Colonial Period, it is possible to reconstruct the socio-political and economic aspects of Inca life. Our knowledge of their general ethnobotany and of their agricultural economy in particular is more satisfactory than for any of the earlier cultures of the Central Andean area. Lists of both the wild and the cultivated plants that were used are available.

Those species that had been domesticated in previous epochs continued as as important cultigens. No new cultivated species were added until the Spanish introduced European species which they brought with them to the New World. The agricultural practices of subjugated areas on the coast continued to be used, although these were often perfected in order to obtain the maximum production of crops. In the highlands, agricultural techniques reached their peak of development with the construction of the impressive terracing and irrigation systems, the remains of which are monuments to the imagination, skill and enterprise of the Inca in the development of their agricultural economy. The ensuing Colonial Period saw the collapse of the most highly organized and complex agricultural system attained in the prehistoric New World.

EXPLANATION OF PLATES

Plant Remains

PLATE I. *Roots, Rhizomes and Tubers (with starch grains)* 1. *Oxalis tuberosa* Mol. Small oca tuber (Botanical Museum 5–376). Pachacamac. 2. *Solanum* sp. Potato (Botanical Museum 5–377) Pachacamac. 3. *Pachyrrhizus tuberosus* (Lam.) Spreng. Roots of jicama (Peabody Museum 30–4190) Paracas. 4. *Manihot esculenta* Crantz. Roots of yuca (Botanical Museum a and b 8–546; c 8–522). Cahuachi. 5. *Ipomoea Batatas* (Linn.) Poir. Sweet potato root (American Museum of Natural History 41.0–5699). Paracas. 6. *Canna* sp. Rhizome of achira (Botanical Museum 5–345). Pachacamac.

PLATE II. *Legumes* 1. *Arachis hypogaea* Linn. Peanuts. a (Botanical Museum 8–21), Chavin-Vasuralis, South Coast; b and c (Botanical Museum 8–611), Cahuachi; d and e (Botanical Museum 7–4), Puerto de Supe, Central Coast. 2. *Phaseolus vulgaris* Linn. Pod of the common bean (Botanical Museum 5–266), Pachacamac; a, red bean (Botanical Museum 8–505), Cahuachi; b and c, two views of a black bean (American Museum of Natural History B–1016), Ancón. 3. *Erythrina* sp. Pods (Botanical Museum a, 5–264; b. 5–255) and seeds (Botanical Museum 5–264). Pachamac. 4. *Phaseolus lunatus* Linn. a, lima bean pod and seed (Botanical Museum 8–627), Cahuachi; b and c, two views of a black lima bean (American Museum of Natural History B–1032), Ancón. 5. *Inga Feuillei* DC. Pacae pod and seeds (Botanical Museum 5–230). Pachacamac. 6. *Canavalia* sp. a, red seed (Botanical Museum 8–615), Cahuachi; b and c, two views of a black seed (Botanical Museum 8–615), Cahuachi; d, pod and seed (Botanical Museum 5–240), Pachacamac.

PLATE III. *Fruits and Seeds* 1. *Bunchosia armeniaca* (Cav.) Rich. a and b, two views of a seed (Botanical Museum 5–278) c, a small seed (Botanical Museum 5–275). Both specimens from Pachacamac. 2. *Persea americana* Mill. Two cotyledons of avocado (Botanical Museum 6–33). Castillo de Tomaval, Virú Valley. 3. *Psidium Guajava* Linn. Guava fruits. a (Botanical Museum 8–378) and b (Botanical Museum 8–554), Cahuachi; c. fruit broken longitudinally (Botanical Museum 6–31), Castillo de Tomaval, Virú Valley; d. half of a fruit (Botanical Museum 8–597), Cahuachi; e, seeds from fruit (Botanical Museum 6–31), Castillo de Tomaval, Virú Valley. 4. *Campomanesia lineatifolia* Ruiz et Pav. Fruits. a and b (Botanical Museum 8–483), Cahuachi; c and d, half of a fruit and seeds (Botanical Museum 372), from the same site; e, cut fruit with seeds removed (Botanical Museum 5–282), Pachacamac. 5. *Lucuma bifera* Linn. a, whole lucuma fruit (Botanical Museum 6–1), Huaca de la Cruz, Virú Valley; b, large seed (Botanical Museum 6–1) from the same site; c, end of a smaller seed (Botanical Museum 5–338), Pachacamac; d. cotyledons (Botanical Museum 5–300), Pachacamac; e, half of a fruit with seeds removed (Peabody Museum 24090).

PLATE IV. *Miscellaneous* 1. *Mucuna* sp. Seed on a cotton string (Botanical Museum 5–229). Pachacamac. 2. *Lagenaria siceraria* (Mol.) Stand. Two groups of gourd seeds: a (Botanical Museum 7–13), Puerto de Supe, Central Coast; b (Botanical Museum 6–25), Huaca de la Cruz, Virú Valley. 3. *Cucurbita moschata* Duch. a, squash stem (Botanical Museum 6–46), Castillo de Tomaval, Virú Valley; b, seeds (Botanical Museum 5–469). Pachacamac. 4. *Cucurbita maxima* Duch. Seeds (Botanical Museum 8–339), tips reconstructed in drawing. Cahuachi. 5. *Erythroxylon Coca* Lam. Coca leaves and fruits (Botanical Museum 10–124). Vista Alegre,

Rimac Valley. 6. *Sapindus Saponaria* Linn. a and b, fruit and seed (Botanical Museum 5–291), Pachacamac; c and d, fruit and broken seed (Botanical Museum 8–603), Cahuachi. 7. *Capsicum* sp. a and b, pepper pods (Botanical Museum 5–214), Pachacamac; *Capsicum sinense* Jacq. c, pod (Botanical Museum 5–209), Pachacamac. 8. *Nectandra* sp. Two cotyledons on a cotton string (Botanical Museum 2–1). Chancay Valley.

PLATE V. *Maize Ears: types of prehistoric maize showing variations in size, row arrangement and kernel characteristics.* 1. Enclosed in husk (Botanical Museum 8–259) with marked venation. Cahuachi. 2. Small, rounded ear (Peabody Museum 46–77–28), rows spiral to the right, kernels prominently oblong. Specimens 2 and 3 among those found in a string bag. Nazca. 3. Rows arranged in cross-spirals, kernels pointed (Peabody Museum 46–77–28). Nazca. 4. Upper half of ear (Peabody Museum 8774 Bag C) consists of straight rows of oblong kernels; lower half demonstrates cross-spiralling of beaked and semi-beaked kernels. Some kernels slightly dented. From string bag. Ancón. 5. Slightly tapered ear (Peabody Museum 8774 Bag C) of oblong kernels arranged in parallel rows which spiral to the left. Ancón. 6. Ear (Peabody Museum 8774 Bag C) with straight rows of predominantly oblong kernels with pronounced horizontal denting. Also from the string bag in which 4 and 5 were recovered. Ancón. 7. Slightly tapered ear (Peabody Museum 8845 Bag A) with beaked kernels arranged in rows primarily spiralling to the left. Ancón.

Reproductions of Plant Forms

PLATE VI. A. Molded warty squash, length approximately 33 cm. (Peabody Museum 30–1658). Coastal Tiahuanaco. B. Two jars in the form of squashes (American Museum of Natural History).

PLATE VII. A. Drinking tube with ear of maize forming the handle; fronds of *Polypodium* sp. decorate the small aryballo (British Museum 1947 AM 10). Inca. B. Stirrup jar with ears of maize forming the body of the vessel. (Museo National, Lima). Mochica.

PLATE VIII. A. Bowl containing three molded ears of maize (Museo National, Lima). Mochica. B. "Peanut Woman," height 6⅜ inches (University Museum 39–20–33, Philadelphia). Mochica.

PLATE IX. A. Stirrup jar decorated with molded peanuts (American Museum of Natural History). B. Pacae pod (Boston Museum of Fine Arts 79.5). Late Chimu.

PLATE X. A. Jar in the form of the rhizomes of achira (Peabody Museum 40–5403). B. Potato jar, height 6¾ inches (University Museum 33591, Philadelphia). Huaca Pariaso.

PLATE XI. A. "Potato Mother," height 10⅛ inches (University Museum 39.20.32, Philadelphia). Mochica. B. Jar with bridge between the two spouts missing; bowls of the squat, adjoining vessels resemble potatoes with pronounced "eyes"; figure of a monkey holding an ear of maize sits on top of the spout of one jar (University Museum 31121, Philadelphia). Pachacamac.

PLATE XII. Stirrup jar decorated with molded fruits of *Bunchosia*, height 20.5 cm. (Peabody Museum 1317). Late Chimu.

PLATE XIII. A. Jar with pepino fruits on the shoulder of the vessel (British Museum 2786). Pacasmayo. B. Double spout jar in the form of a lucuma fruit (American Museum of Natural History 41.0–884). Nazca.

PLATE XIV. Double spout jar decorated with a border of peppers, height 20 cm. (Peabody Museum A 2670). Nazca.

PLATE XV. A. Mold-made jar of Fanged Diety among maize ears, height 123 cm. (American Museum of Natural History B/8863). Mochica. B. Jar representing tubers of Tunta (Chicago Museum of Natural History Cat. No. 1188). Chimbote, Mochica style.

PLATE I

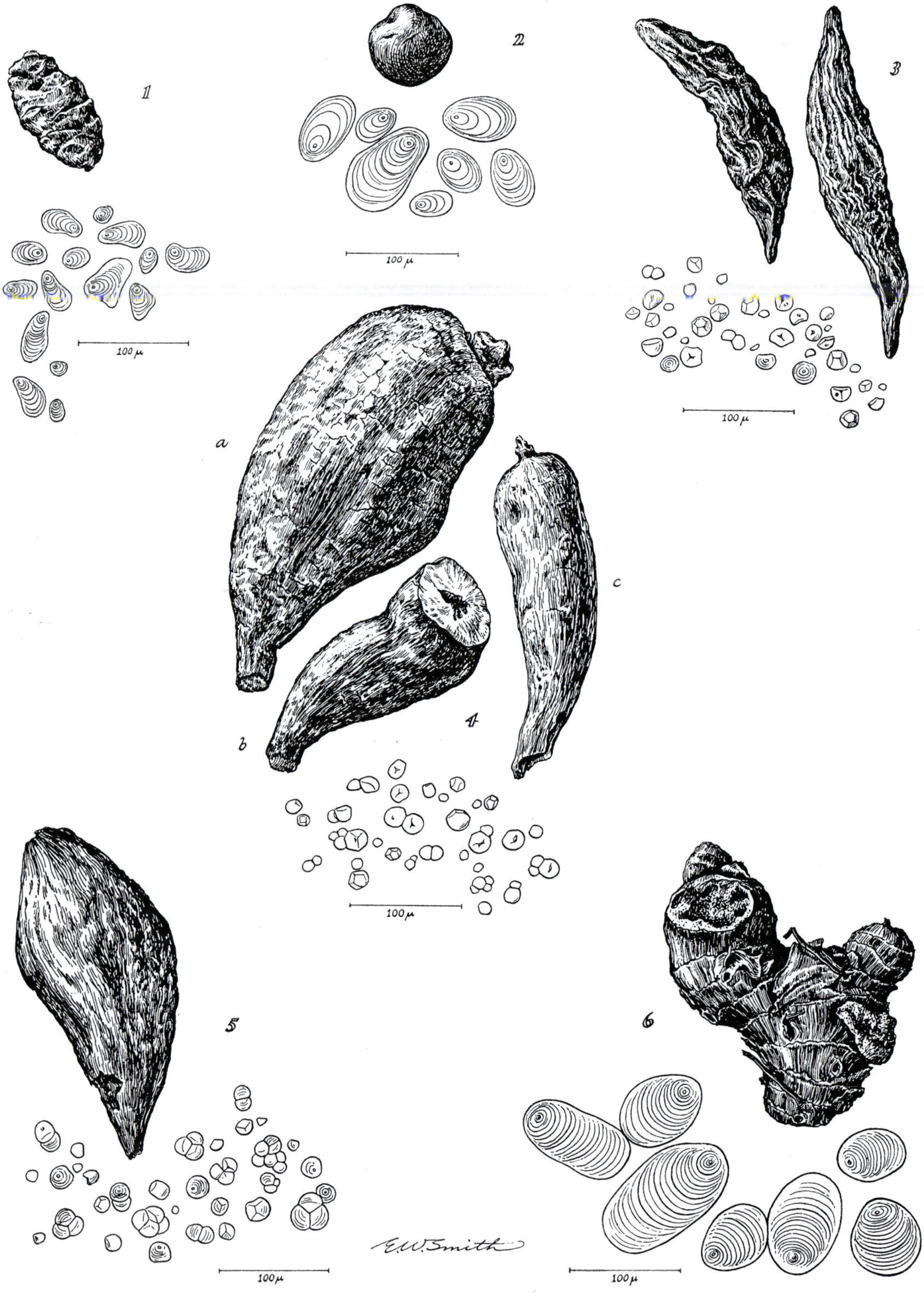

Plate II

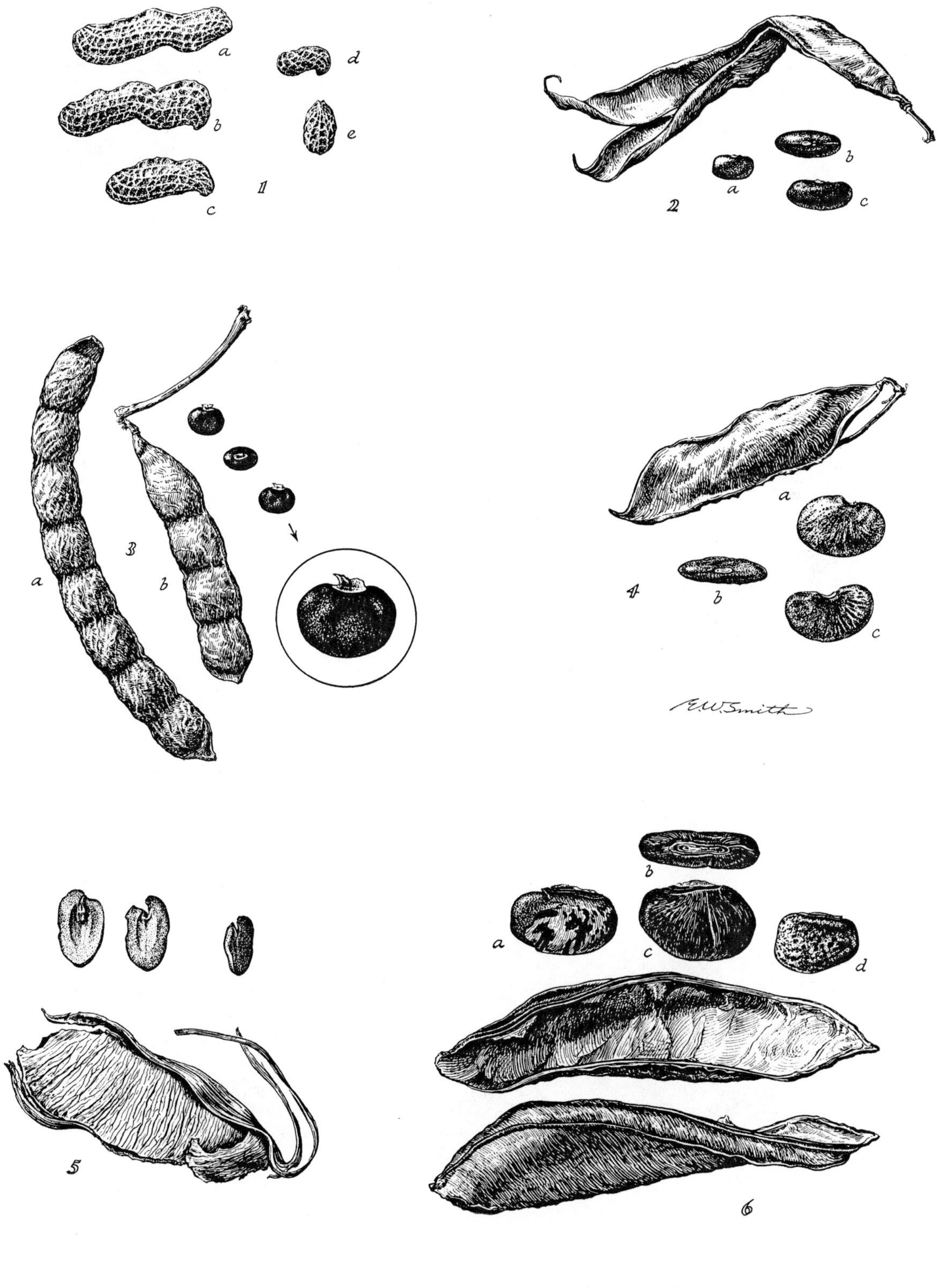

Plate III

PLATE IV

Plate V

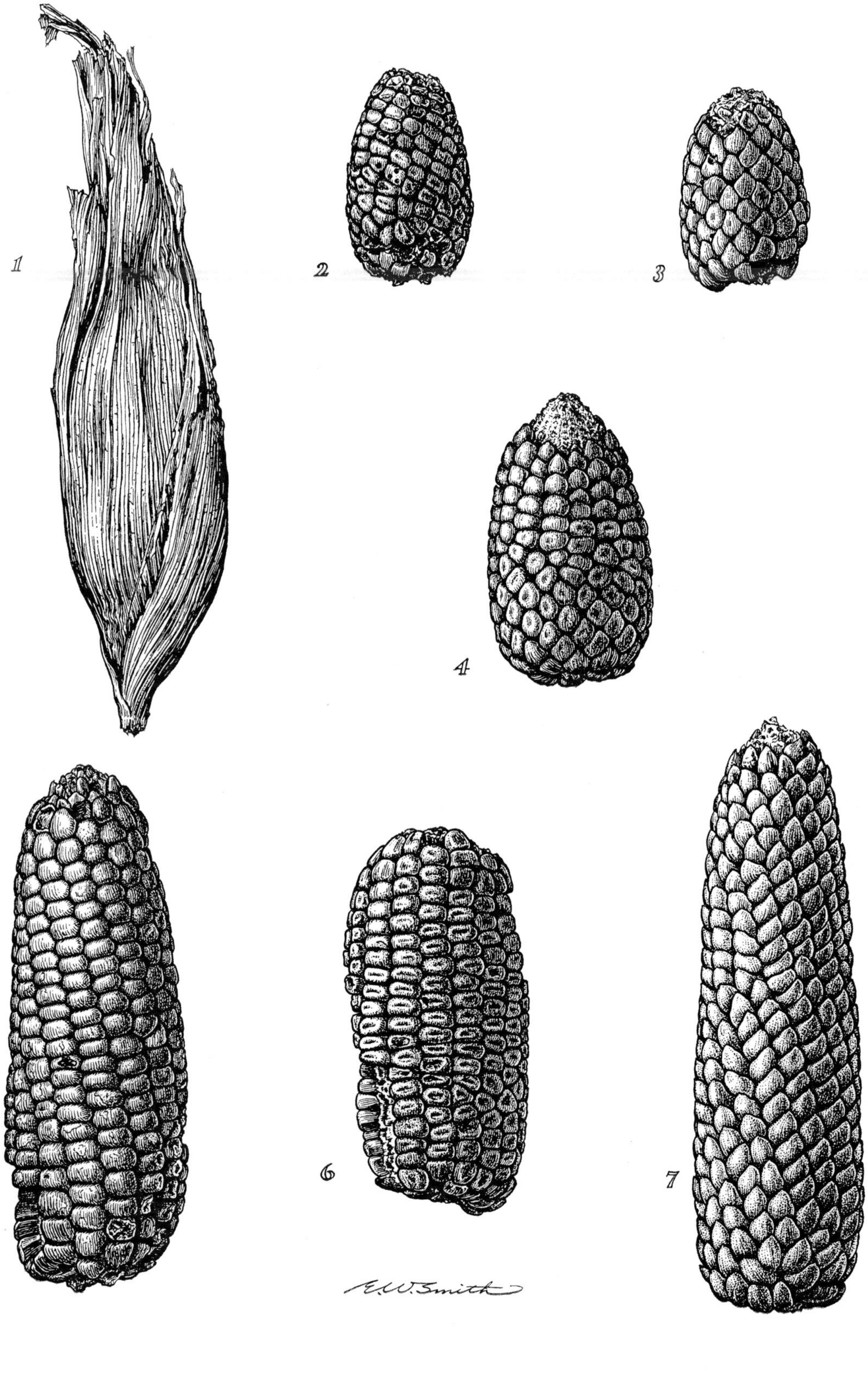

PLATE VI

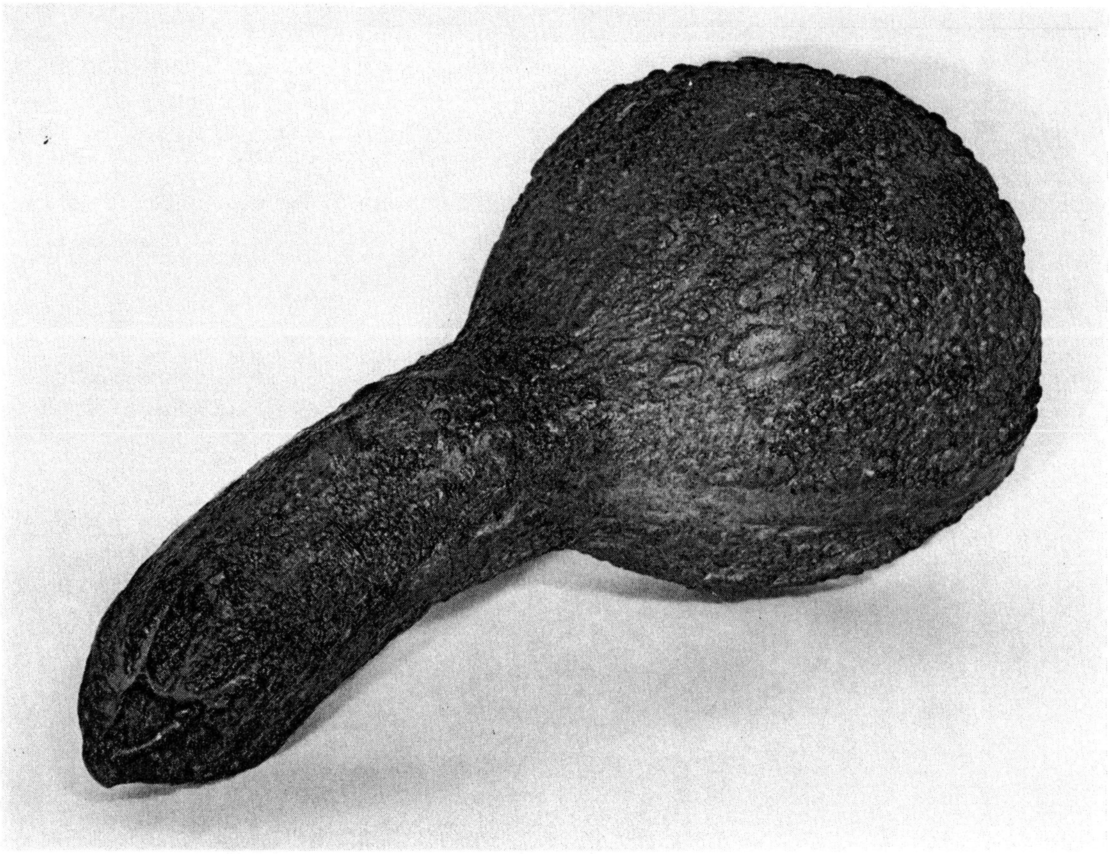

A

B

Plate VII

A

B

PLATE VIII

A

B

Plate IX

A

B

Plate X

A

B

Plate XI

A

B

PLATE XII

Plate XIII

A

B

PLATE XIV

PLATE XV

A

B

BIBLIOGRAPHY

Ames, Oakes

1939 *Economic Annuals and Human Cultures.* Botanical Museum, Harvard University. Cambridge.

Anderson, Edgar E.

Report on the maize recovered at Arica, Chile. (MS).

Anonymous

1938 "If You Died In Old Peru," *Natural History,* 41:119–25. New York.

Arbeláez, Enrique Pérez

1956 *Plantas Utiles de Colombia.* Vol. I. Bogota.

Bailey, L. H. et al.

1917 *The Standard Cyclopedia of Horticulture.* New York.

Beetle, Alan A.

1945 "Sedge-Boats in the Andes," *Journal of the New York Botanical Garden,* 46:1–4. New York.

Bennett, Wendell C.

1946a "The Andean Highlands: An Introduction." In *Handbook of South American Indians,* Bureau of American Ethnology, Bulletin 143, 2:1–60. Washington.

1946b "The Archaeology of the Central Andes." *Ibid.,* pp. 61–147.

1948 "The Peruvian Co-Tradition." In "A Reappraisal of Peruvian Archaeology," *American Antiquity,* 13:107. Menasha, Wisconsin.

Bennett, Wendell C. and Bird, Junius B.

1949 *Andean Culture History.* American Museum of Natural History Handbook Series No. 15. New York.

Bird, Junius B.

1943 "Excavations in Northern Chile." *Anthropological Papers,* American Museum Natural History, 38:171–318. New York.

1946 "The Cultural Sequence of the North Chilean Coast." In *Handbook of South American Indians,* Bureau of American Ethnology Bulletin 143, 2:587–594. Washington.

1948a "America's Oldest Farmers," *Natural History,* 57:296–303; 334–335. New York.

1948b "Preceramic Cultures in Chicama and Virú." In "A Reappraisal of Peruvian Archaeology," *American Antiquity,* 13:21–8. Menasha, Wisconsin.

1951 "South American Radiocarbon Dates," *American Antiquity,* 17:37–48. Menasha, Wisconsin.

Brown, C. Barrington

1926 "On Stone Implements from North-West Peru," *Man,* 26:97–101. London.

Bukasov, S. M.

1930 *The Cultivated Plants of Mexico, Guatemala and Colombia.* Bulletin of Applied Botany, Genetics and Plant-breeding, Supplement 47. Leningrad.

1933 *The Potatoes of South America and their breeding possibilities.* Bulletin of Applied Botany, Genetics, and Plant-breeding, 58th Supplement. Leningrad.

CARDENAS, MARTIN

1948 *Plantas Alimenticias Nativas de los Andes de Bolivia.* Folia Universitaria, No. 2. Cochabamba, Bolivia.

CARRION CACHOT, REBECA

1948 "La Cultura Chavin. Dos nuevas colonias: Kuntur Wasi y Ancon," *Revista del Museo Nacional Antropologia y Arquelogia,* 2:99–172. Lima.

1949 *Paracas; Cultural Elements.* Lima.

CARTER, GEORGE F.

1945 "Some Archaeological Cucurbit Seeds from Peru," *Acta Americana,* 3:163–172. Cambridge.

CLAUSEN, ROBERT T.

1945 *A Botanical Study of the Yam Beans* (Pachyrrhizus). Cornell University Agricultural Experiment Station, Memoir 264. Ithaca.

COBO, BERNABE

1890–5 *Historia del Nuevo Mundo.* 4 vols. Seville.

COLLIER, DONALD

1955a "Cultural Chronology and Changes as Reflected in the Ceramics of the Virú Valley, Peru," *Fieldiana: Anthropology,* Vol. 43. Chicago.

1955b "Development of Civilization on the Coast of Peru." *Social Science Monographs 1,* Pan-American Union, pp. 19–27. Washington.

COOK, O. F.

1910 "History of the Coconut Palm in America," *Contribution, U.S. National Herbarium,* 14:271–342. Washington.

1925 "Peru As a Center of Domestication," *Journal of Heredity,* 16:33–46, 95–110. Washington.

CORRELL, DONOVAN S.

1952 *Section Tuberarium of the Genus* Solanum *of North America and Central America.* U.S. Department of Agriculture, Agricultural Monograph 11. Washington.

COSTANTIN, J. and BOIS, D.

1910 "Sur les graines et tubercules des Tombeaux péruviens de la Periode Incasique," *Revue Générale de Botanique,* 22:242–265. Paris.

CUTLER, HUGH C.

1946 "Races of Maize in South America," *Botanical Museum Leaflets,* 12:257–291. Cambridge.

CUTLER, HUGH C. and CARDENAS, MARTIN

1947 "Chicha, A Native South American Beer," *Ibid.,* 13:33–60.

DAHLGREN, B. E.

1936 *Index to American Palms.* Field Museum of Natural History, Publication 355. Chicago.

DEWEY, LYSTER H.

1943 *Fiber Production in the Western Hemisphere.* U.S. Department of Agriculture, Miscellaneous Publications No. 518. Washington.

ENGEL, FREDERIC

1956 "Curayacu–A Chavinoid Site," *Archaeology,* 9:98–105. Cambridge.

1957a "Early Sites on the Peruvian Coast," *Southwest Journal of Anthropology*, 13:54–68. Albuquerque.

1957b "Sites et Établissements sans Ceramique de la Côte Peruvienne," *Journal de la Sociéte des Américainistes*, n.s. 46:67–156. Paris.

ESTETE, MIGUEL DE

1918 "N(oticia) del Perú." In *El descubrimiento y la conquista del Peru.* Sociedad ecuatoriana de estudios historicos Americanos, 1:300–50. Quito.

FESTER, G. A.

1953 "Einige Farbstoffe sudamerikanischer Kulturvolker," *Isis*, 44:13–16. Cambridge.

FESTER, GUSTAVO A. and CRUELLAS, JOSÉ

1934 "Colorantes de Paracas," *Revista del Museo Nacional*, 3:154–156. Lima.

FORD, JAMES ALFRED and WILLEY, GORDON R.

1949 *Surface Survey of the Virú Valley, Peru.* American Museum of Natural History, Anthropology Papers, 43:1–90. New York.

GAYTON, A. H. and KROEBER, A. L.

1927 *The Uhle Pottery Collection from Nazca.* University of California Publications in American Archaeology and Ethnology, 24:1–46. Berkeley.

GROBMAN, ALEXANDER, SALHUANA, WILFREDO, and MANGELSDORF, PAUL C.

1956 *Races of Maize in Peru.* Escuela Nacional de Agricultura, Maize Genetics Co-operation, News Letter 30. Lima.

HARMS, HERMAN VON

1922 "Ubersicht der bisher in altperuanishen Grabern gefundenen Pflanzenreste," in *Festschrift Eduard Seler*, 157–86. Stuttgart.

HARSHBERGER, J. W.

1898 "Uses of Plants among the Ancient Peruvians," *Bulletin of the Museum of Science and Art*, 1:1–4. University of Pennsylvania. Philadelphia.

HAWKES, J. G.

1947 "On the Origin and Meaning of South American Potato Names," *Journal of the Linnean Society of London*, 53. Botany No. 350:205–50. London.

HEISER, CHARLES B. and SMITH, PAUL G.

1953 "The Cultivated Capsicum Peppers," *Economic Botany*, 5:214–227. New York.

HERRERA, FORTUNATO L.

1921 *Contribucion a la Flora del Departmento del Cuzco.* Primera Parte, Segunda Edicion. Cuzco.

1923 "Fitolatria Indigena. Plantas y Flores simbolicos de los Inkas," *Inca*, 1:440–446. Lima.

1939 *Catalogo Alfabetico de los Nombres Vulgares y Cientificos de Plantas que Existen en el Peru.* Lima.

1940 "La Quihuicha no es el *Amaranthus caudatus* como impropiamente la denomina el Professor O. F. Cook," *Revista Museo Nacional*, 9:229–235. Lima.

1941a "Plantas Alimenticias Domesticadas por los Inkos cuyo Uso se Encuentra poco Generalizado," *Chaski*, 1:11–18. Lima.

1941b *Sinopsis de la Flora del Cuzco.* Tomo I. Lima.

1942a "Plantas Endemicas Domesticades por los Antiguos Peruanos," *Revista Museo Nacional*, 11:25–30. Lima.

1942b "Plantas tropicales cultivados pos antiguos peruanos," *Ibid.*, 11:179–195.

HILL, ALBERT F.

1952 *Economic Botany.* 2nd ed. New York.

HODGE, W. H.

1946 "Three Neglected Andean Tubers," *New York Botanical Garden,* 47:214–224. New York.

1954 "The Edible Arracacha," *Economic Botany,* 8:195–221. New York.

HORKHEIMER, HANS

1958 *La Alimentación en el Peru Prehispanico y su interdependencia con la Agriculture.* UNESCO, Programa de Estudios de la Zona Arida Peruana. Lima.

HUNZIKER, ARMANDO T.

1943 *Granos Hallados en el Yacimiento Arqueologica de Pampa Grande (Salta, Argentina).* Buenos Aires.

HUTCHINSON, J. B., SILOW, R. A. and STEPHENS, S. G.

1947 *The Evolution of Gossypium and the Differentiation of the Cultivated Cottons.* Oxford.

HUTCHINSON, SIR JOSEPH

1959 *The Application of Genetics to Cotton Improvement.* Cambridge.

JAMES, PRESTON E.

1942 *Latin America.* New York.

JIJON Y CAAMAÑO, JACINTO

1949 *Maranga: contribucion al conacimento de los aborigenes del Valle del Rimac,* Peru. Quito.

KAPLAN, LAWRENCE

1956 "The Cultivated Beans of the Prehistoric Southwest," *Annals of the Missouri Botanical Garden,* 43:189–251. St. Louis.

KIDDER, ALFRED II

1956 "Digging in the Titicaca Basin," *University Museum Bulletin,* 20:17–29. Philadelphia.

KOSTERMAN, A. J. G. H.

1936–38 "Revision of the Lauraceae," *Mededeelingen van het Botanisch Museum en Herbarium van Rijks Universiteit te Utrecht,* No. 33–50. Utrecht.

KRAPOVICKAS, ANTONIO and RIGONI, VICTOR A.

1957 "Nuevas Especies de "Arachis" vinculades al Problema del Origen del Mani," *Darwiniana,* 11:431–455. San Isidro, Argentina.

KROEBER, ALFRED L.

1944 *Peruvian Archaeology in 1942.* Viking Fund Publication in Anthropology, Number 4. New York.

1948 *Anthropology.* New York.

KROEBER, ALFRED L. and STRONG, WILLIAM DUNCAN

1924a "The Uhle Collection from Chincha," University of California Publications in American Archaeology and Ethnology, 21:57–94. Berkeley.

1924b "The Uhle Pottery Collection from Ica," *Ibid,* 21:126–133.

LABARRE, WESTON

1947 "Potato Taxonomy Among the Aymara Indians of Bolivia," *Acta Americana,* 5:83–103. Cambridge.

1948 "The Aymara Indians of the Lake Titicaca Plateau, Bolivia," *American Anthropologist,* 50:1–250. Menasha, Wisconsin.

LARCO, HOYLE, RAFAEL

1938, 1939 *Los Mochicas.* Vols. I and II. Lima.

1946 "A Culture Sequence for the North Coast of Peru." In *Handbook of South American Indians,* Bureau of American Ethnology, Bulletin 143, 2:149–175. Washington.

LEVI-STRAUSS, CLAUDE

1950 "The Use of Wild Plants in Tropical South America." *Ibid,* 6:165–186.

MACBRIDE, J. FRANCIS

1936–56 *Flora of Peru.* Field Museum of Natural History, Botanical Series, 13. Chicago.

MACKIE, W. W.

1943 "Origin, Dispersal and Variability of the Lima Bean, *Phaseolus lunatus,*" *Hilgardia,* 15:1–24. Berkeley.

MANGELSDORF, PAUL C.

1942 Identifications of Vegetal Remains from Nazca and Paracas. (MS). Cambridge.

MANGELSDORF, PAUL C. and OLIVER, DOUGLAS L.

1951 "Whence Came Maize to Asia?" *Botanical Museum Leaflets,* 14:263–291. Harvard University. Cambridge.

MARKHAM, CLEMENTS ROBERT

1864 *Contributions toward a Grammar and Dictionary of Quichua.* London.

MASON, J. ALDEN

1957 *The Ancient Civilization of Peru.* Harmondsworth, Middlesex, England.

MEJÍA XESSPE, M. T.

1931 "Kausay (Alimentacion de los Indios)," *Wira-Kocha,* 1:9–24. Lima.

NORDENSKIOLD, ERLAND

1931 "Origin of the Indian Civilizations of South America." In *Comparative Ethnographical Studies,* No. 9:1–76. Goteborg.

O'NEAL, LILA M. and WHITAKER, THOMAS W.

1947 "Embroideries of the Early Nazca Period and the Crop Plants Depicted on Them," *Southwest Journal of Anthropology,* 3:294–321. Albuquerque.

PARODI, LORENZO R.

1926 "Contribution à l'étude des plantes alimentaires indigenes cultivées en Argentine," *Revue de Botanique Appl.,* 16:177–189. Paris.

1932 "(Las) Balsas usa dos por los aymara en el largo Titicaca," *Physics,* 11:145–149 (*Revista de la Sociedad Argentina de Ciencias Naturales*). Buenos Aires.

1935 "Relaciones de la Agricultura Prehispanica con la Agricultura Argentina Actual," *Anales de la Academia Nacional de Agronomia y Veterinaria de Buenos Aires,* 1:115–167. Buenos Aires.

PICKERING, CHARLES

1879 *Chronological History of Plants.* Boston.

POPENOE, WILSON and JIMÉNEZ, OTON

1921 "The Pejibage: a Neglected Food-Plant of Tropical America," *Journal of Heredity,* 12:154–166. Washington.

RADIN, PAUL

1942 *Indians of South America.* New York.

RECORD, SAMUEL J. and HESS, ROBERT W.

1943 *Timbers of the New World.* New Haven.

REISS, WILHELM and STÜBEL, ALPHONS

1880–87 *The Necropolis of Ancón in Peru.* 3 Vols. New York, London, and Berlin.

ROCHEBRUNE, ALPHONSE TREMEAU DE

1879 "Recherches d'ethnographie botanique sur la flore des sépultures Péruviennes d'Ancon," *Actes Societe Linneaus Bordeaux,* 3:343–358. Bordeaux.

ROWE, JOHN HOWLAND

1946 "Inca Culture." In *Handbook of South American Indians.* Bureau of American Ethnology, Bulletin 143, 2:183–330. Washington.

SAFFORD, WILLIAM EDWIN

1917a "Food-Plants and Textiles of Ancient America," in *Proceedings of the 19th International Congress of Americanists,* pp. 12–30. Washington.

1917b "A Forgotten Cereal of Ancient America." *Ibid.,* pp. 286–97.

1926 "The Potato of Romance and Reality." In *Smithsonian Report 1925,* pp. 509–32. Washington.

SAFFRAY, DR.

1876 "Les antiquities péruviennes a l'exposition de Philadelphia," *La Nature,* 4:401–407. Paris.

SALAMAN, REDCLIFFE N.

1949 *The History and Social Influence of the Potato.* Cambridge.

SAUER, CARL O.

1950 "Cultivated Plants of South and Central America." In *Handbook of South American Indians.* Bureau of American Ethnology, Bulletin 143, 6:487–543. Washington.

SAUER, JONATHAN D.

1950 "The Grain Amaranths: A Survey of Their History and Classification," *Annals of the Missouri Botanical Garden,* 37:561–632. St. Louis.

SCHAEDEL, RICHARD P.

1951 "Major Ceremonial and Population Centers in Northern Peru." In *Twenty-ninth International Congress of Americanists, pp.* 1:232–242. Chicago.

1957 "Highlights of Andean Archaeology 1954-1956," *Archaeology,* 10:93–99. Brattleboro, Vermont.

SMITH, PAUL G. and HEISER, CHARLES B., JR.

1957 "Taxonomy of *Capsicum sinense* Jacq. and the Geographic Distribution of the Cultivated *Capsicum* species," *Bulletin of the Torrey Botanical Club,* 84:413–420. New York.

STANDLEY, PAUL C.

1917 "North American Flora," *New York Botanical Garden,* 21:95–169. New York.

1930 *Flora of Yucatan.* Field Museum of Natural History, Publ. 279. Chicago.

STRONG, WILLIAM DUNCAN

1925 "The Uhle Pottery Collection from Ancon," *University of California Publications in American Archaeology and Ethnology,* 21:135–190. Berkeley.

1943 *Cross-Section of New World Prehistory: A Brief Report on the Work of the Institute of Andean Research. 1941–42.* Smithsonian Miscellaneous Collection, 104: No. 2. Washington, D.C.

1948 "Cultural Epochs and Refuse Stratigraphy in Peruvian Archaeology," *American Antiquity,* 13:93–102. Menasha, Wisconsin.

1954 "Recent Archaeological Discoveries in South Coastal Peru," *Transactions of the New York Academy of Sciences*, Ser. II, 16:215–218. New York.

1957 "Paracas, Nazca and Tiahuanacoid Cultural Relationships in South Coastal Peru," *American Antiquity*, 22:1–48. Salt Lake City.

STRONG, WILLIAM DUNCAN and CORBETT, JOHN M.

1943 "A Ceramic Sequence at Pachacamac." In *Archaeological Studies in Peru, 1941–42*, pp. 27–121. New York.

STRONG, WILLIAM DUNCAN and WILLEY, GORDON R.

1943 "Archaeological Notes on the Central Coast." *Ibid.* pp. 5–25.

STRONG, WILLIAM DUNCAN, WILLEY, GORDON R. and CORBETT, JOHN M.

1943 *Archaeological Studies in Peru, 1941–42*. New York.

STRONG, WILLIAM DUNCAN and EVANS, CLIFFORD, JR.

1952 *Cultural Stratigraphy in the Virú Valley, Northern Peru.* New York.

STUMER, LOUIS M.

1954a "The Chillón Valley of Peru: Excavation and Reconnaissance 1952–1953. Part 1," *Archaeology*, 7:171–78. Brattleboro, Vermont.

1954b "The Chillón Valley of Peru: Excavation and Reconnaissance 1952–1953. Part 2," *Archaeology*, 7:220–28. Brattleboro, Vermont.

1954c "Population Centers of the Rimac Valley in Peru," *American Antiquity*, 22:130–148. Salt Lake City.

1954d "Report on the South Peruvian Coast: Chala to Arica," *American Antiquity*. 19:384–86. Salt Lake City, Utah.

TELLO, JULIO C.

1929 *Antiguo Peru: primera epoca.* Comision Organizadora del Segundo Congreso Sudamericano de Turismo. Lima.

1930 "Andean Civilization: Some Problems of Peruvian Archaeology." In *Proceedings of the 23rd International Congress of Americanists*, pp. 259–290. New York.

1938 *Arte Antiguo Peruana. Inca.* Vol. 2. Lima.

TOWLE, MARGARET A.

1948 Report on the Plant Remains from Pachacamac (MS). Cambridge.

1952a "The Pre-Columbian Occurrence of Lagenaria Seeds in Coastal Peru," *Botanical Museum Leaflets*, 15:171–184. Harvard University. Cambridge.

1952b "Plant Remains from a Peruvian Mummy Bundle," *Ibid.*, pp. 223–246.

1952c "Descriptions and Identifications of the Virú Plant Remains." In Strong, Wm. Duncan and Evans, Clifford, Jr.: *Cultural Stratigraphy in the Virú Valley, Northern Peru*, pp. 352–56. New York.

1954 "Plant Remains." In Willey, Gordon R. and Corbett, John M.: *Early Ancón and Early Supe Culture*, pp. 130–38. New York.

1956 Descriptions and Identifications of Plant Materials from the South Coast of Peru (MS). Cambridge.

1957 Plant Remains from the Site of Chiripa, Bolivia (MS). Cambridge.

1958 Plant Remains from Certain Sites in the Rimac Valley, Peru (MS). Cambridge.

TRYON, ALICE F.

1959 "Ferns of the Incas," *American Fern Journal*, 49:10–24. Lancaster, Pa.

TSCHOPIK, HARRY, JR.

1946 "Some Notes on Rock Shelter Sites near Huancayo, Peru," *American Antiquity*, 12:73–80. Menasha, Wisconsin.

UHLE, MAX

1903 *Pachacamac*. Philadelphia.

1913 "Die Muschelhugel von Ancon, Peru." In *International Congress of Americanists*, Session 18, pp. 22–45. London.

1914 "The Nazca Pottery of Ancient Peru." In *Proceedings of the Davenport Academy of Sciences*, 13:1–16. Davenport, Iowa.

1924a *See* Kroeber and Strong: *The Uhle Pottery Collection from Ica.*

1924b "Ancient Civilization of Ica Valley," *University of California Publications in American Archaeology and Ethnology*, 21:128–133. Berkeley.

1925 "Report on Exploration at Supe," *University of California Publications in American Archaeology and Ethnology*, 21:257–264. Berkeley.

U.S. DEPARTMENT OF AGRICULTURE

1917 *New Plant Introductions, 6th Annual List 1916–1917*. Washington.

1918 "Inventory of Seeds and Plants Imported by Office of Plant Introduction during the Period from October 1 to December 31, 1915." Bureau of Plants, *Introduction Inventory* 45 (1918) 15. Washington.

VARGAS, CESAR

1947 "Algunas fitomorfias raras o criticas de los antiguos peruanos," *Boletin de la Sociedad Argentina de Botanica*, 2:39–42. La Plata.

WEBERBAUER, A.

1945 *El Mundo Vegetal de Los Andes Peruanos. Estudio Fitogeografico.* Estación Experimental Agricola de la Molina Direccion de Agricultura Ministerio de Agricultura. Lima.

WHITAKER, THOMAS W.

1948 "Lagenaria: a Pre-Columbian Cultivated Plant in the Americas," *Southwestern Journal of Anthropology*, 4:49–68. Albuquerque.

WHITAKER, THOMAS W. and BIRD, JUNIUS B.

1949 "Identification and Significance of the Cucurbit Materials from Huaca Prieta, Peru." *American Museum Novitates*, 1426. New York.

WHITAKER, THOMAS W. and BOHN, G. W.

1950 "The Systematic, Genetic, Production and Uses of the Cultivated Cucurbits," *Economic Botany*, 4:52–81. New York.

WIENER, CHARLES

1880 *Perou et Bolivie*. Paris.

WILLEY, GORDON R.

1943 "A Supplement to the Pottery Sequence at Ancon." In *Archaeological Studies in Peru, 1941–42*. New York.

1951 "Peruvian Settlement and Socio-Economic Patterns." In *The Civilization of Ancient America. Selected Papers of the XXIXth International Congress of Americanists*, 195–200. Chicago.

1953 *Prehistoric Settlement Patterns in the Virú Valley, Peru*. Bureau of American Ethnology, Bulletin 155. Washington.

1955 "The Prehistoric Civilizations of Nuclear America," *American Anthropologist*, 57:571–593.

WILLEY, GORDON R. and CORBETT, JOHN M.

1954 *Early Ancon and Early Supe Culture. Chavín Horizon Sites of the Central Peruvian Coast.* New York.

WILLEY, GORDON R. and PHILLIPS, PHILIP

1955 "Methods and Theory in American Archaeology, II: Historical-Developmental Interpretation," *American Anthropologist,* 57:723–819.

WILLIAMS, LLEWELLYN

1936 *Woods of Northeastern Peru.* Field Museum of Natural History Botanical Series 15. Chicago.

WITTMACK, L.

1880–87 "Plants and Fruits." In Reiss, Wilhelm and Stübel, Alphons: *Necropolis of Ancón in Peru,* Vol. 3, Pt. 13. New York, London, and Berlin.

1888 "Die Nutzpflanzen der alten Peruanes," in *Congrès International de Americanistes,* 7:325–349. Berlin.

YACOVLEFF, E.

1933 "La jiquima, raiz comestible extinguida en el Peru," *Revista del Museo Nacional,* 3:51–66. Lima.

YACOVLEFF, E. and MUELLE, J. C.

1932 "Una Exploracion en Cerro Colorado," *Ibid.,* 1:31–103.

1934 "Un fardo funerario de Paracas," *Ibid.,* 3:63–153.

YACOVLEFF, E. and HERRERA, F. L.

1934–35 "El Mundo vegetal de los antiguos peruanos," *Ibid.,* 3:241–322; 4:29–102.

INDEX

CPSIA information can be obtained at www.ICGtesting.com
Printed in the USA
268389BV00005B/37-50/P